# LEWIS and CLARK

## IN MISSOURI

# LEWIS and CLARK

## IN MISSOURI

THIRD EDITION ——————— ANN ROGERS

UNIVERSITY OF MISSOURI PRESS
COLUMBIA AND LONDON

Library of Congress Cataloging-in-Publication Data

Rogers, Ann.
    Lewis and Clark in Missouri / Ann Rogers.—3rd ed.
        p.  cm.
    Includes bibliographical references and index.
    ISBN 0-8262-1413-4 (alk. paper).—ISBN 0-8262-1415-0 (pbk. : alk. paper)
        1. Lewis and Clark Expedition (1804–1806).  2. Lewis, Meriwether, 1774–
1809.  3. Clark, William, 1770–1838.  4. Missouri—Description and travel.
5. Historic sites—Missouri.  I. Title.
F592.7 .R63  2002
917.804'2—dc21                                                    2002022564

Design: Jennifer Cropp
Composition: Stephanie Foley
Printer and binder: International Printing Services (IPS)
Typefaces: Minion and Palatino

This book has been published with the generous assistance of a contribution
from the William T. Kemper Foundation, Commerce Bank, Kansas City,
Missouri.

*FOR MY FATHER AND FOR JOE*

# Contents

# PREFACE

Like most people fortunate enough to have been caught up in the story of Lewis and Clark, my husband and I began our retracings of the expedition's route in the Northwest. We took the boat trip through the Gates of the Mountains, drove the narrow road that crosses the Continental Divide at Lemhi Pass, and visited the reconstruction of Fort Clatsop, set in a forest of coastal Oregon. Although we are both Missourians, we at first neglected Missouri's role in our national epic of exploration.

The story of that great journey is, of course, so multifaceted that it can never be captured in a single retelling, and any writer or photographer must choose to emphasize some aspects at the expense of others. Anyone familiar with books on the Lewis and Clark Expedition knows that the Missouri portion of the story is one consistently abbreviated or even ignored. Often a few pages or a single photograph will carry the Corps of Discovery from the base camp at Wood River, Illinois, to the Mandan villages in North Dakota.

Such slight attention to Missouri is regrettable because its chapter in the story of Lewis and Clark is pictorially beautiful and historically significant. Missouri was the setting for five months of preparation by Meriwether Lewis. The journey began with a ten-week crossing of Missouri and ended more than two years later with the explorers' triumphant arrival at the St. Louis riverfront. Missouri was the place many members of the expedition, including the captains, chose to live after their return.

In the following pages the often-told story is told with an emphasis on the Missouri chapter, one that holds interest for Missourians and all who follow Lewis and Clark.

# ACKNOWLEDGMENTS

The deepest roots of this book are in my father's love of the American West. My interest in photographing its landmarks and learning its history began on trips with him.

Thirty years ago my husband suggested we retrace a portion of the Lewis and Clark route, and we have made almost annual journeys since, following various segments of the trail between Charlottesville and Astoria.

Winifred George, who would later become president of the Lewis and Clark Trail Heritage Foundation, introduced me to that organization. The foundation publishes *We Proceeded On,* with articles on specific aspects of the expedition, and holds annual meetings at different locations along the trail. I have benefited from the articles and meetings and from friendships formed with those who share a common interest.

Jerry Garrett has planned many enjoyable, substantive meetings of the Metro–St. Louis Chapter and has generously given additional time to reading the manuscript for this book. I am grateful for his insights and comments.

Mimi and Darold Jackson have brought seemingly endless energy to making the Lewis and Clark Center and the Discovery Expedition of St. Charles effective tools for teaching, and I have always found them ready to answer a question or assist in other ways on my Lewis and Clark projects.

Joanne and Glen Bishop have been an inspiration to all. Glen provided a wonderful symbol for the Missouri portion of the expedition when he handcrafted a full-scale replica of Lewis and Clark's keelboat, and he provided an amazing example of courage when he accepted the loss of that boat in a warehouse fire and set out to replace it by building another. The Discovery Expedition of St. Charles now has full-size replicas of the three boats used in the 1804 journey across Missouri.

In writing this book, I have been fortunate to have at hand Gary Moulton's edition of the Lewis and Clark journals. My task was lightened by having all of the journals and related material in one edition, with updated notes, and by having these volumes on my bookshelf and readily available.

Lewis and Clark material in the form of letters and other documents continues to come to light. James Holmberg, curator of special collections at the Filson Historical Society in Louisville, has written about the recently discovered Clark letters and on the fate of Lewis's dog. I find the Clark letters regarding Lewis's death and the story of Seaman's collar the most fascinating of the previously unknown "sequels" to the expedition, and I am glad I could weave some of this material into the current edition.

It was an honor to be named to the Missouri Lewis and Clark Bicentennial Commission by Governor Mel Carnahan. Serving on the commission has given me a greater appreciation for the work done by the Missouri Department of Conservation and the Department of Natural Resources. I am grateful to Shannon Cave for his assistance and support and to Jim Denny for the information and assistance he has provided. I am especially grateful to Jonathan Kemper, cochair of the commission, for his interest in *Lewis and Clark in Missouri* and his support for a new edition.

The St. Louis Camera Club has been part of my life for many years, a legacy of my father's interest in photography. Its members enjoy competition but are remarkably willing to help one another. Photographs by two members, Lynda Stair and Casey Galvin, appear in this book.

The Missouri Historical Society's library is a valuable resource for anyone researching the history of the American West and especially of Missouri, and I am fortunate to have its collection so easily available. I thank its staff for the assistance provided.

Frances Stadler, the society's former archivist, was a personal friend, and shortly before her death she presented me with several Lewis and Clark books from her own library. I treasure them and my memories of her.

I am grateful to the staff of the Missouri State Archives for providing requested material and to members of the St. Louis County Library's reference department, who responded graciously to numerous requests for information.

I am also grateful to Barbara Tettaton, an illustrator on the staff of the St. Louis County Library, who prepared the maps used in this edition.

Dave Bennett, a historian at Fort Osage, sent me material regarding Joseph Whitehouse's service there. I thank him and all who sent information, answered questions, and provided photographs and other illustrations.

I thank Connie Wakefield, who taught me to use a computer, and Alan Gorman, a computer consultant, who repeatedly rescued me when I ranged beyond my skills.

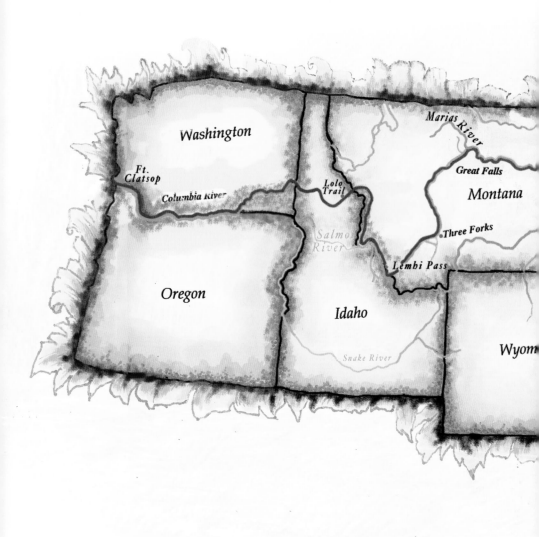

St. Joseph

Kansas River
Fort Osage
Kansas City

Rocheport

St. Charles

St. Louis

Jefferson City

Mississippi River

Cape Girardeau

Missouri

# LEWIS and CLARK

## IN MISSOURI

Ezra Winter's mural entitled *St. Louis: The Way Opened to the Pacific* depicts the 1804 ceremony transferring the Louisiana Territory to the United States. *State Historical Society of Missouri, Columbia.*

# 1

## PREPARING FOR THE JOURNEY

M eriwether Lewis and William Clark were in St. Louis when the
flags of three nations flew over the town within a period of
twenty-four hours. March 9, 1804, was the date set for the ceremony
marking the transfer to the United States of the French-owned and
Spanish-administered Louisiana Territory. While Spanish troops stood
at attention in front of the government house, Spain's flag was lowered and replaced with the flag of France. As a courtesy to the many
French residents of the town, Captain Amos Stoddard, the principal
U.S. representative, then allowed the French tricolor to fly for a day
before his soldiers raised the Stars and Stripes, signaling America's
possession of the Louisiana Territory. Drum rolls, gun salutes, and
speeches accompanied the changing of the colors. Stoddard, who
accepted the territory in the name of the United States, invited Captain
Lewis to step forward and sign the transfer document as the chief official witness.

Lewis had arrived in St. Louis three months earlier, carrying a pass-
port that identified him as a "Citizen of the United States, who, by
authority of the President . . . is setting out on a voyage of discovery
with the purpose of exploring the Missouri river and the western
regions of the Northern Continent."[1]

The man chosen for this mission was born in Virginia, on August 18,
1774, about nine miles from Jefferson's home at Monticello. He was
five years old when his father, a soldier in the Revolutionary War, died
of pneumonia contracted after he crossed a flooded stream while
returning to duty following a brief winter visit with his wife and children. As a young boy, Lewis developed a taste for solitary adventure
that would remain an integral part of his character. He seemed most at
home roaming fields and forests as a hunter, fisherman, woodsman,
and naturalist.

He first learned to hunt in his native Albemarle County, then
improved his wilderness skills when his mother remarried and his
stepfather moved the family from Virginia to a frontier settlement in

Georgia. When Lewis was about thirteen, he returned to Virginia to receive a few years of schooling and to help manage Locust Hill, the plantation he had inherited from his father. He quit his studies at eighteen, deciding to concentrate solely on life as a gentleman farmer, but by twenty he grew restless and enlisted in the Virginia militia. After several promotions he transferred to the regular army and served for a time in a rifle company commanded by William Clark, who would later share command of their expedition.

In March 1801, twenty-six-year-old Captain Lewis received a letter from Thomas Jefferson, a family friend. The newly elected president asked him to become his secretary and aide, an invitation Lewis accepted. Historians disagree on whether Jefferson was already considering him to lead a western expedition, but Lewis's two years at the White House and Monticello gave him a background in science, diplomacy, Indian relations, and other fields that would later serve him in carrying out the president's plans for an American reconnaissance of the Northwest.

The idea of an exploration of the western half of the continent had been taking shape in Thomas Jefferson's mind for at least twenty years before Lewis and Clark arrived in the St. Louis area to complete preparations for their "voyage of discovery." In 1783, while a delegate to the Continental Congress and a member of the American Philosophical Society, Jefferson had asked George Rogers Clark, one of William Clark's older brothers, if he would lead a western expedition. The Revolutionary War hero, by then physically and financially exhausted, declined his friend's invitation. Two years later, while serving as minister to France, Jefferson encouraged an American adventurer named John Ledyard, who had the idea of crossing the continent from west to east. Ledyard didn't make the planned journey; and in 1793, Jefferson, still a member of the American Philosophical Society and by then Secretary of State, wrote detailed instructions to a French botanist, André Michaux. But Michaux became entangled in political intrigues, and his trip ended before he reached the Mississippi.

Jefferson's own travels never took him more than fifty miles west of Monticello, but he wanted to learn whatever he could about the lands beyond the Mississippi: the major rivers and their tributaries, the native peoples and their languages, the plants and animals, the geography and geology, the climate and terrain. He foresaw the eventual expansion of the United States to the Pacific and believed an American presence at the mouth of the Columbia River would strengthen claims to that region. He found additional incentive for an American expedition through the Northwest when the account was published in 1801 of Alexander Mackenzie's overland journey through Canada in search of a water route to the Pacific. Mackenzie hadn't found the northwest passage he sought, but Jefferson believed one existed and that it offered the promise of facilitating American trade with the Far East.

In February 1803, he wrote letters to Benjamin Barton, Benjamin Rush, and Caspar Wistar, telling the Philadelphia scientists that Congress had secretly approved his plan to send a party of ten or twelve men, led by Meriwether Lewis, up the Missouri River and then on to the Pacific. Explaining that Lewis would be arriving in their city within a few weeks, he asked each man to assist in preparing him for his mission. Lewis would spend about a month in Philadelphia, learning ways to gather the diverse information Jefferson wanted and ways to keep his men sound throughout the long journey. Benjamin Rush, the leading physician of his day, recommended specific medicines and provided Lewis with his "Rules of Health." Benjamin Barton, a botanist as well as physician, perhaps discussed with his guest the best ways to preserve and label the plants he would be collecting. Barton loaned Lewis his copy of Du Pratz's *History of Louisiana,* and Lewis in turn purchased a copy of Barton's textbook, *Elements of Botany.* He carried both on the expedition and found Barton's of more use. Caspar Wistar, a physician and paleontologist who shared Jefferson's fascination with mammoths and mastodons, probably gave Lewis instruction in finding ice-age fossils along the bluffs of the Missouri River. In Lancaster, Pennsylvania, en route to Philadelphia, Lewis had met with astronomer and surveyor Andrew Ellicott, who responded to Jefferson's request that he teach Lewis to take celestial observations. That instruction continued in Philadelphia with Robert Patterson, a mathematician and another member of the American Philosophical Society. Lewis acquired the books and navigational instruments Ellicott and Patterson recommended, but he never acquired a high level of skill for determining latitude and longitude.[2]

Along with learning, Lewis was busy spending some of the twenty-five hundred dollars appropriated by Congress on a massive but carefully considered array of goods. Purchases for his men included tents, camping equipment, fishing gear, knives, tools such as saws and chisels, writing instruments, blankets, coats, and shirts, including the flannel shirts recommended by Dr. Rush. Gifts for the Indians included fishhooks, needles, tobacco, beads of various colors, mirrors, scissors, cloth, ribbons, handkerchiefs, lockets, rings, and broaches. To carry out his scientific observations, he purchased a chronometer, sextants, compasses, and other instruments used for navigation and surveying. He also bought gunpowder, powder horns, and flints for rifles and muskets. Guns and a substantial amount of ammunition would be needed for the hunters to supply adequate game.

Before coming to Philadelphia, he had visited the United States armory at Harpers Ferry, in present-day West Virginia, where he ordered fifteen Model 1803 rifles, the finest available. With these, his men would be able to make an impressive show of strength if challenged and repel an attack if necessary.

July brought two pieces of welcome news that helped define the expedition. Early in the month, the president told Lewis he had received confirmation of the Louisiana Purchase. U.S. representatives in Paris had been trying to assure access to the port of New Orleans when they learned Napoleon was willing to sell not only that city but also about 820,000 square miles west of the Mississippi. The fifteen-million-dollar purchase nearly doubled the size of the United States and extended the nation as far west as the Rocky Mountains. There was no longer a need for the journey to be secret; in fact, one of its new goals was to acquaint the Indians living in the Louisiana Territory with the United States government and its interest in trade.

Later that month, a letter from William Clark further defined the expedition. Lewis had written on June 19 to his friend and former commander, outlining plans for the trip and inviting Clark "to participate . . . in it's fatiegues, it's dangers and it's honors," honors that would include, he was told, a captain's commission, as authorized by the president. One day after receiving Lewis's letter, Clark responded that while he realized the difficulties and dangers inherent in the venture, "no man lives whith whome I would perfur to undertake Such a Trip . . . as your self."[3]

The Lewis and Clark partnership would prove to be one of mutual respect and superb effectiveness. The backgrounds of the two men were similar; their abilities, it is often pointed out, were complementary. William Clark, four years older than Lewis, had also been born in Virginia, and his family, like Lewis's, had lived in Albemarle County. Clark was the ninth of ten children and the youngest of six sons. When he was fourteen, he moved with his family to Kentucky, a region George Rogers Clark had come to know during his military campaigns against the British. William Clark joined the militia, then transferred to

Statues of Meriwether Lewis and William Clark by James Earle Fraser in the Missouri capitol at Jefferson City portray Clark as the expedition's cartographer and Lewis as its visionary leader. *Ann Rogers.*

the regular army, serving in the Indiana and Ohio Territories under General Anthony Wayne.

As much at home in the wilderness as Lewis, Clark had a greater interest in Indians and a better understanding of them. Complementing the zeal and determination of the somewhat solitary Lewis, he was practical, steady, and at ease with others. He was interested in natural history, and his military training may have included instruction in designing and constructing forts, as well as in cartography, a skill he would demonstrate to a remarkable degree as he mapped the expedition's route to the Pacific.

By the time Clark's letter of acceptance reached him, Lewis had moved on to Pittsburgh, where he spent six frustrating weeks waiting for the keelboat he had ordered, a delay due, he wrote, "to the unpardonable negligence and inattention of the boat-builders."[4] When finally completed, the boat was fifty-five feet long and about eight feet wide at its center, with a thirty-two-foot mast that would allow its crew to take advantage of favorable breezes. Before leaving Pittsburgh on August 31, Lewis purchased a smaller craft, called a pirogue, and welcomed a pilot, two or three prospective recruits, and seven soldiers sent from a post near Philadelphia, who together were expected to get the boats down the Ohio.

His frustrations continued on the river, which was unusually low, even for late summer. Riffles were common, repeatedly forcing the crew to unload the keelboat and attempt to lift or drag it past the obstructions. When manpower failed, Lewis hired horses or oxen from nearby farms, although he complained he was being overcharged. Many stretches of river were without any perceptible current or movement except what the wind created, and because the winds blew most frequently from the west, they offered little help to those traveling downstream. Some of the crew deserted or were discharged. Along the way, he purchased a second pirogue and a series of canoes, most of which leaked and had to be replaced.

By October 14, Lewis had reached the Falls of the Ohio, where the river dropped over a series of limestone ledges, forming two miles of rapids. On the south bank, at the eastern end of these falls, was Louisville, where William Clark had lived with his parents. On the north bank, below the falls, was Clarksville, where he was living in 1803 with George Rogers Clark. Because Lewis broke off his journal after his entry for September 18 and did not resume writing until November 11, he provided no description of his reunion with Clark or his first impressions of other men who would play significant roles in the expedition.[5]

Lewis's letter had asked Clark to look for some healthy young men who were "good hunters . . . accustomed to the woods, and capable of bearing bodily fatigue in a pretty considerable degree." They could be told the journey would involve a voyage up the Mississippi and an absence of perhaps eighteen months, but Lewis didn't want the "real design" disclosed until he was ready to engage them.[6] In response, Clark had assembled a number of possible candidates, seven of whom were inducted into the army along with George Shannon and John Colter, who had come down the Ohio with Lewis. The seven were William Bratton, George Gibson, John Shields, brothers Reubin and Joseph Field, and cousins Charles Floyd and Nathaniel Pryor. Together with Colter and Shannon, they formed the nucleus of the Corps of Discovery and have become known as the "nine young men from Kentucky." Also joining the party at this point was York, a slave Clark had inherited under the terms of his father's will. The two had been companions since boyhood, and York would travel with Clark to the Pacific.

After almost two weeks in the Louisville area, Meriwether Lewis, joined now by William Clark and the growing number of recruits, continued down the Ohio toward its juncture with the Mississippi. About thirty-five miles east of the convergence, near present-day Metropolis, Illinois, they reached Fort Massac, where Lewis was able to hire interpreter George Drouillard. The son of a French Canadian father and Shawnee mother, Drouillard knew a number of Indian languages, including sign language, and was an excellent hunter and pathfinder, all skills that would serve the expedition well. He would be paid monthly, since he had not yet agreed to make the entire journey. Drouillard's first assignment was to travel about three hundred miles to South West Point, in Tennessee, and bring any recruits the commander released to meet Lewis and Clark's party, which would be heading north on the Mississippi.

Lewis discharged at Fort Massac the soldiers who had helped bring his boats down from Pittsburgh, so he reached the Mississippi with about a dozen men, all of whom were expected to be members of the permanent party. Also on board and expected to be part of the permanent party was his Newfoundland dog, Seaman, described by Lewis as "very active strong and docile." Newfoundlands, which weigh 110 to 150 pounds when mature, were, in Lewis's time, working dogs. Besides being used to pull carts on land, they were often used by fishermen to help haul in nets or for rescues because the large, powerful dogs are intelligent and thoroughly at ease in the water. The first mention in Lewis's journal of his Newfoundland tells how the dog would kill squirrels in the river and "swiming bring them in his mouth to the boat." (Lewis "thought them when fryed a pleasent food.")

In the months ahead Seaman would demonstrate other abilities. He could bring down a wounded deer or kill an antelope by drowning it in the river. When the expedition moved through grizzly bear country, Seaman's barking would alert the explorers to the presence of

Lewis described his Newfoundland
as "active strong and docile." The
dog proved to be a valuable member
of the expedition. *Ann Rogers.*

those dreaded animals. One frightening night in 1805, a bison charged
past men lying beside their campfire and then headed toward the tent
where the captains were sleeping. When the intruder had sped off
into the darkness and the startled party discovered no one had been
harmed, Lewis credited Seaman, reporting: "My dog saved us by causing him to change his course."[7]

Long before that night in eastern Montana, Lewis put a high value
on his dog. The twenty dollars he paid for the Newfoundland was
equal to half his month's salary as aide and secretary to President
Jefferson. He again showed the value he placed on Seaman when he
refused to sell him. The refusal was, in fact, the first recorded action of
Meriwether Lewis in the future state of Missouri. On November 16,
two days after reaching the confluence of the Ohio and Mississippi,
the captains crossed to the west bank of the Mississippi, where they
found some Delawares and Shawnees encamped. Lewis wrote of the
meeting that took place in what is now Mississippi County, Missouri:
"One of the Shawnees a respectable looking Indian offered me three
beverskins for my dog with which he appeared much pleased, the dog
was of the newfoundland breed one that I prised much for his docili-
ty and qualifications generally for my journey and of course there was
no bargan." It was in the Northwest, eighteen months later, that Seaman
would best demonstrate his "qualifications." But even on the banks of

the Mississippi in November of 1803, Lewis knew there was no gain for him in trading his dog.

Much of the week spent at the confluence was given over to the initial use of the brass sextant, chronometer, and other instruments for measurement that Lewis had acquired in Philadelphia. The captains' attempts to determine latitude and longitude would be repeated, with varying degrees of success, at major confluences and other important sites throughout the long journey. Clark, who had learned the basics of surveying, calculated the widths of these two rivers and then drew a map of their convergence to include in his notes.

This was not William Clark's first visit to the area that would become Missouri. In the autumn of 1795, he had been sent by General Anthony Wayne on a mission that combined rhetoric with reconnaissance. The young lieutenant, accompanied by seventeen soldiers, traveled down the Ohio and crossed the Mississippi to New Madrid, where he protested to Spanish officials the building of a Spanish fortification on the east side of the Mississippi. On his return from the meeting, he visited the remains of Fort Jefferson, established by George Rogers Clark and named for Thomas Jefferson, who had been governor of Virginia when the fort was built. Located a few miles south of the confluence, it had been abandoned after only two years, following repeated attacks by Indians.[8]

In 1802, a year before William Clark saw the Fort Jefferson site with Meriwether Lewis, his older brother had written to Jefferson, telling the president that when the secretary of war had requested information about the confluence area, with a view toward reestablishing a fort, William had provided a complete description. George Rogers Clark added: "If it should be in your power to confur on him any post of Honor and profit, in this Countrey in which we live, it will exceedingly gratify me."[9] When Jefferson and Lewis discussed Clark's role in the expedition, his brother's request may have carried at least some weight.

As the captains viewed the site, their thoughts were already turned to the future. Knowing the lands across the Mississippi would be formally transferred to the United States within a few months, they had traveled a short distance down the opposite bank two days earlier to ascertain the best location for a fort on the west side of the confluence area.

On November 20, 1803, Lewis and Clark began their ascent of the Mississippi. For the next three weeks, they and their party would see the landscape of the future state of Missouri as it unfolded to them along the west side of the river. During these weeks, the two leaders would be taking observations, recruiting additional men, making journal entries, and developing the partnership that would play a major role in the success of their expedition. The great journey was still ahead, but the weeks on the Mississippi form an important preface to that epic.

Although the land to the west was under the control of the Spanish government, the Corps of Discovery's boats soon began passing American settlements on Tywappity Bottom, an extensive plain located in what would become Scott and Mississippi Counties. Rich soil, coupled with generous Spanish land grants, had induced farmers from the United States to immigrate to this fertile region years in advance of the Louisiana Purchase.

About forty miles north of the confluence, on November 23, Lewis came ashore at the Spanish post of Cape Girardeau. Carrying a letter of introduction from Daniel Bissell, the officer in charge at Fort Massac, he paid a visit to Louis Lorimier, the commandant. A French Canadian who had been a trader among the Indians in Ohio, Lorimier sided with the British during the Revolutionary War and, with some help from his Indian allies, captured Daniel Boone. In return, George Rogers Clark attacked and burned Lorimier's trading post.

A few years later, Lorimier crossed the Mississippi, settling first along Saline Creek, south of Ste. Genevieve, and then moving to Cape Girardeau, where he established a successful trade with emigrant Shawnees, Delawares, and members of other eastern tribes who had also moved across the river. Encouraged by Lorimier to come to the region, these peaceful Indians were welcomed by the Spanish officials as a buffer between white settlers and the feared Osage tribe to the west.

The Cape Girardeau trading post known as "Lorimier's Red House" was his home, as well as the seat of civil and military authority for the district, and it was here that Lewis dined with the commandant and his family. In spite of his Tory background, Lorimier impressed his American guest with his striking appearance and considerable charm. Lewis's journal describes him as a robust man of about sixty, with dramatically dark eyes and thick black hair, which he wore in a braid hanging "nearly as low as his knees." He told Lewis it had once been long enough to touch the ground when he stood.

His wife, Charlotte Bougainville, was part Shawnee, and her manner of dress combined the French style with Indian leggings and moccasins. Lewis surmised she had been "very handsome when young." The couple's large family included a daughter Lewis described as

Cape Rock, the site of the original settlement at Cape Girardeau, is shown in a painting by Humphrey Woolrych. *University Museum at Southeast Missouri State University, Cape Girardeau, Missouri; gift of Robert and Richard N. Remfro.*

"an agreeable, affible girl, & much the most descent looking feemale I have seen since I left the settlement in Kentuckey a little below Louisville." After dinner, one of Lorimier's sons accompanied Lewis on the three-mile trip by horseback to Old Cape Girardeau, where Clark had brought the boats to shore for the night. The site, now known as Cape Rock, is where Jean Girardot first visited as a young French ensign and returned to establish a trading post. The name "Cape Girardot" appeared on maps of the region as early as 1765.[10]

From Lorimier, or perhaps from Drouillard, or perhaps from the pilot aboard his boat, Lewis had learned of settlements currently in the area, and he noted two in his journal. About sixteen miles west of Cape Girardeau was a large community of "duch" (actually German) descendants, a "temperate, laborious and honest people," who had "erected two grist mills and a saw-mill." A few miles to the north was the Apple River, later called Apple Creek. Along this Missouri stream, some seven miles from its entrance into the Mississippi, was a Shawnee village, where as many as four hundred people lived by hunting and farming.[11]

The legendary Grand Tower, now called Tower Rock, is a landmark on the Mississippi River.
*Casey Galvin.*

Shortly before sunset on November 25, Lewis and Clark arrived at the limestone formation they knew as the Grand Tower, an insular rock near the west bank of the Mississippi. As the crew made camp, Lewis used the remaining daylight to scale the tower. Once on top, he dropped a cord from the southeast corner to measure its height, which he recorded as 92 feet. The surrounding waters posed no unusual risk that afternoon, but Lewis was aware that when the river was high "an immence and dangerous whirlpool" formed as a second channel forced its way between the tower and a nearby formation before rejoining the main channel. If any boats dared to approach under those conditions,

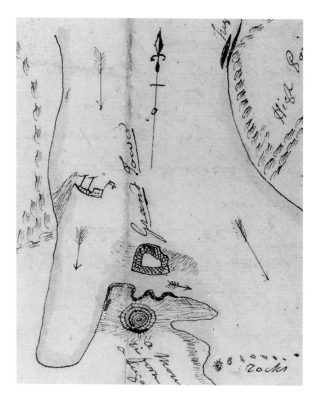

While Lewis climbed the Grand Tower, Clark drew a map of the area and included a small sketch of the keelboat.
*Yale Collection of Western Americana, Beinecke Rare Book and Manuscript Library.*

Lewis wrote, "the counter courent . . . would instantly dash them to attoms and the whirlpool would as quickly take them to the botom."

Now known as Tower Rock, the monolith was an often described feature of the Mississippi and the subject of numerous legends. Lewis had picked up some of the lore and wrote that the tower had the same symbolism for rivermen as the equator has for deepwater sailors: "Those who have never passed it before are always compelled to pay or furnish some sperits to drink or be ducked." An earlier traveler familiar with the landmark was Father Jacques Marquette, who had seen the tower more than a century before with Louis Joliet. Marquette's journal describes the "furious combat" of the waters and "a great roaring" that led Indians in the region to believe "there is a manitou there, that is, a demon who devours all who pass."[12]

Lewis observed that the tower seemed to be a vestige of hills that once extended eastward across the Mississippi but "in the course of time have been broken down by the river." And, from the summit of the nearby mound in the form of a sugarloaf, he enjoyed what he described as "a most beautifull and commanding view."

While Lewis explored and wrote, Clark mapped the area. The *Atlas of the Lewis and Clark Expedition*, edited by Gary E. Moulton, includes two maps Clark made of the site, including one on which he sketched, numbered, and described ten geographic features, beginning with the "Grand Tower in the Mississippi." To both of these maps Clark added a small drawing of the keelboat, to mark the expedition's campsite on the west bank of the river, just north of the tower. The drawings show the boat as having two masts, each with a yard. A flag flies from a staff mounted on the windowed cabin at the boat's stern. These sketches are less detailed and far less familiar than the pair of drawings he made months later as the vessel was being readied for its 1804 voyage. Those give a deck plan showing the placement of oars and a profile showing the boat with a single mast.[13] But the tiny sketches on the Grand Tower maps are Clark's earliest representations of the keelboat in which the Corps of Discovery would travel sixteen hundred miles up the Missouri River.

Two days beyond the Grand Tower, Clark assumed command of the keelboat and crew while Lewis visited Kaskaskia and Cahokia, on the east side of the Mississippi, recruiting more men, obtaining supplies, and making other preparations related to the expedition. Lewis would also cross the river to discuss with Spanish officials in St. Louis his planned ascent of the Missouri. On the day he left the keelboat, he turned over to Clark the duty of journal-keeping and apparently did not resume making regular entries until April 1805, when the expedition left the Mandan villages in present-day North Dakota.[14]

On the November morning William Clark took over the journal, visibility was poor, and he strained to see more clearly the high bluffs that gave this stretch of the river what he called "a bold and rockey shore." By midafternoon the sky turned dark and the current grew increasingly swift as the crew threaded a dangerous course past sandbars before bringing the keelboat to shore for the night. A few miles below, on the west side of the river, was a landing place for boats carrying salt from several nearby licks to the village of Ste. Genevieve. Salt for trade, rich soil for agriculture, and the availability of lead ore had given the town a measure of prosperity and a population that for a time rivaled that of St. Louis.

Clark made no mention of visiting the village on this journey, but he had seen it on his second trip west of the Mississippi, in 1797, while trying to untangle his older brother's financial affairs. (George Rogers Clark had led a small band of men that captured British out-

The rebuilt gatehouse at Fort de Chartres is part of a reconstruction of the
French fort Clark saw in ruins in 1803. *Ann Rogers.*

posts in Illinois during the Revolutionary War, but his later years were
plagued by debts and lawsuits.) On William Clark's 1797 visit to Ste.
Genevieve, a town then "Situated on the Spurs of the high land," he
had been the guest of François Vallé II, the commandant.[15] Clark's stopping point in 1803 was opposite old Ste. Genevieve, the site of the original settlement, which suffered repeated flooding and was abandoned
after the devastating flood of 1785.

About fifteen miles to the north of the village, Clark found another
reminder of his earlier journey through this area. On the Illinois side of
the Mississippi stood the remains of Fort de Chartres. From an island
opposite the fort, he took readings to try to determine the latitude of
the site and recorded in his journal that he had been able to see two
sides of the structure, which was then in ruins. As a military man, he
knew Fort de Chartres had been the pride of French America, the most
impressive fortification on this part of the continent. The stone structure built in 1756 replaced earlier wooden stockades eroded by flood-
waters. Massive limestone walls eighteen feet high and nearly four feet
thick enclosed four acres of facilities capable of housing four hundred

men. It was so much a symbol of French dominion in America that when France ceded the lands east of the Mississippi to England in 1763, Fort de Chartres was the last remnant of French power to be relinquished. The English used the fort for a time as the seat of government in the Illinois country but within a decade surrendered it to the Mississippi River, the force that had destroyed the two previous stockades.

On Clark's first visit, he was traveling by horseback and was able to ride up to the remains. His 1797 journal describes the parade ground as overgrown with thornbushes. Owls were the fort's inhabitants, and deer came to lick in the rooms where salt had been stored.[16] Six years later his view of the ruins was from a distance, and when he had taken his observations the boat moved on. Above the fort, he turned his attention to the Missouri side of the Mississippi, where the cliffs contained "a number of Indented Arches" of various sizes, which he felt gave a "verry romanteck appearance" to this part of the river.

In what is today Jefferson County, Missouri, Clark noted a stream with several mills built along it and wrote: "Emigrent americains are Settled verry thick up this Creek." Americans had also settled along Joachim Creek, which he recorded in his journal as "Swacken Creek," a spelling not far from the common form, Swashin Creek. Both represent American efforts to reproduce the French pronunciation of the saint's name. Joachim Creek flows into the Mississippi near the town of Herculaneum.[17]

About noon, at the mouth of the River des Peres, the boats passed a village officially named Carondelet, for the Baron de Carondelet, who served as governor of the Louisiana Territory during its administration by Spain. Clark would have recognized the name from his earlier visits to the Mississippi Valley, but he also knew the place as "Vide Poche," or "empty pocket," a nickname bestowed by its more affluent neighbor to the north. In 1803, the village of Carondelet had about fifty homes, inhabited chiefly by persons of French and Spanish descent, along with a number of Indians. Although its setting was often described as idyllic, the community of farmers and voyageurs could not match the wealth of St. Louis, where the fur trade was beginning to create a prosperous merchant class.

That evening Clark brought the keelboat to the Illinois shore opposite St. Louis. Little, if anything, could be seen on a rainy December night, but Clark's first impression of St. Louis was written in the journal he kept during his 1797 trip: "I was Delighted from the ferry with the Situation of this Town, which is on the decline of a hill, commanding a butifull view of the river." During that visit he "found it to be in a Thriveing state," with "a number of Stone houses" built on two streets parallel to the Mississippi.[18] By 1804, the town had about two hundred houses, most of them whitewashed stone buildings, nearly square and one story high, set on quarter-block lots.[19]

While Clark waited on the Illinois side of the river, Lewis made his initial visit to St. Louis in the company of John Hay and Nicholas Jarrot, both of Cahokia, who would serve as his interpreters. When Lewis rejoined the boats on December 9, he told Clark that the Spanish commandant, Carlos Dehault Delassus, was refusing to allow their ascent of the Missouri River without first receiving permission from his superiors. The news was not really a setback. In winter, the Missouri could be clogged with enough ice to make navigation virtually impossible, and before summer the Upper Louisiana Territory would be transferred to the United States.

Clark was told to continue north about eighteen miles to a point nearly opposite the Mississippi's confluence with the Missouri and examine a location that had been recommended to Lewis, perhaps by Jarrot, who considered the land to be his. There, beside a small Illinois stream called Wood River, or river Dubois, the Corps of Discovery would build a winter camp and use the time to prepare for the journey west.

The boats arrived at Wood River accompanied by hail and snow, prompting the men to waste no time in constructing winter quarters. With a site chosen the first day, work began immediately. Clark hired a wagon for hauling logs, and the recruits started building their cabins. Within a week, while snow and sleet fell, the men moved into the unfinished structures, as Clark continued working on the chimney of his own hut. The Corps of Discovery greeted Christmas of 1803 with the discharge of rifles. John Shields, one of the "nine young men from Kentucky," returned to camp with cheese and four pounds of butter that he had been able to purchase from white settlers. But a far better gift to share that Christmas was the news that George Drouillard, whom Lewis had hired at Fort Massac, had agreed to accompany the expedition to the Pacific. In the months ahead his skills as a hunter, scout, and interpreter would prove invaluable.

The presence of game was an important factor in choosing the Wood River site, and Drouillard, whose name often appears in the journals as "Drewyer," soon demonstrated his abilities as a hunter. He returned from one foray with three deer and five turkeys, and a few days later, he killed two more deer and reported seeing several bears.

John Colter and Reubin Field were among those also bringing in turkeys and deer, as well as rabbits and raccoons. In the course of the winter, a number of hunters provided meat, four men made sugar, some caught catfish, and others discovered "Bee Trees" with "great quantities of honey."

Maintaining discipline was difficult, especially when hunting and other activities drew men well beyond the bounds of the camp. Lewis accused some of the recruits of using hunting forays as "a pretext to cover their design of visiting a neighbouring whiskey shop," and Clark ordered at least one person to stop selling liquor to his men, but bouts of drinking followed by fights went on. Two chronic offenders, Hugh Hall and John Collins, would continue to cause problems in the first weeks of the journey, but Collins eventually redeemed himself by becoming one of the expedition's best hunters. The journals record only Drouillard, the Field brothers, and Shields as more successful.[20]

On a day in early January, Clark made a reconnaissance of the area that led him to remnants of an earlier culture. Crossing a prairie, he saw nine mounds forming a circle. Two of the mounds rose about seven feet above the plain, and around them he found "great quantities of Earthen ware & flints." The mounds he described as an "Indian Fortification" or "fortress" are now thought to have been the bases for ceremonial structures built between A.D. 900 and 1300. Only one of the mounds believed to have been seen by Clark in 1804 remains, near Mitchell, Illinois. But others in the same group stand within Cahokia Mounds State Historic Site, a few miles to the south.[21]

Clark broke through the ice on a pond near the mounds and returned to camp cold, wet, and with his shoes frozen to his feet—an experience that left him feeling "verry unwell" for several days. His frequent illnesses during the winter encampment, along with those of Nathaniel Pryor and at least three other men, were treated with home remedies. Captain Lewis, drawing on his knowledge of herbal medicine, prescribed walnut bark for Clark's illness; weeks later, when Pryor became ill, Clark sent Reubin Field to kill a squirrel to make soup.

Lewis offered his prescription during one of his brief and infrequent visits to Wood River. He spent much of the winter of 1803–1804 in St. Louis, where he became a friend of Auguste Chouteau and his younger half-brother, Pierre. In 1763, Auguste had been a boy of fourteen when he watched his stepfather, Pierre Laclède, choose the site for St. Louis.[22] In the years that followed, the brothers became wealthy fur traders and leading citizens of the city. They had prospered while

Pierre Chouteau's home, one of the finest in St. Louis, was Lewis's unofficial residence during the last months of preparation for the journey west. *Missouri Historical Society, St. Louis; Clarence Hoblitzelle, artist.*

St. Louis flew the flags of Spain and France, and their desire for a good relationship with the new government made them eager to assist Meriwether Lewis.

Pierre Chouteau's home became Lewis's unofficial residence during the winter of preparation. From the Chouteaus he learned about the lower Missouri region and its native peoples, drawing on the experience the brothers had as established traders with the Osage Indians. Throughout the winter, Lewis continued outfitting the Corps of Discovery with purchases of gunpowder, bullets, knives, blankets, and trade goods from the Chouteau warehouse. The brothers later recruited French Canadian boatmen to accompany the Corps of Discovery on the first stage of the journey, and during that first stage the Chouteaus would send letters to President Jefferson advising him of the expedition's progress as reported to them by their contacts along the river.[23]

On March 26, 1804, Lewis sent Jefferson some slips of wild plum and Osage orange from his host's garden. Because the plum was a small, thick shrub, Lewis felt it might "form an ornimental and usefull

hedg." The Osage orange, as he explained to Jefferson, had been introduced to St. Louis by Pierre Chouteau, who "obtained the young plants
at the great Osage vilage from an Indian of that nation." These Indians,
Lewis learned, "give an extravigant account of the exquisite odour
of this fruit when it has obtained maturity, which takes place the latter
end of summer, or the begining of Autumn."[24] Because the wood of
the tree is tough and little affected by weather conditions, it was
favored by the Indians for making their bows. To this day, the Osage
orange is sometimes called "bodock," a corruption of the French *bois
d'arc* or "bow wood." Historian Donald Jackson notes that President
Jefferson forwarded the slips to nurseryman Bernard McMahon, who
planted some in front of his Philadelphia store. A row of Osage orange
believed to date from that period still grows at the location.[25] Examples
of Osage orange also grow at the Missouri Botanical Garden in St.
Louis and the University of Virginia, whose campus was designed by
Thomas Jefferson, himself an avid gardener.

In a letter written to the War Department, Lewis recommended a
son of Pierre Chouteau and a son of Charles Gratiot, Pierre's brother-
in-law, be given appointments to the recently established United States
Military Academy at West Point. (Lewis was grateful for Charles
Gratiot's knowledge of languages because the Chouteaus were not
fluent in English and Lewis did not speak French.) Pascal Bouis, the
son of another St. Louis merchant, would also be recommended for the
new academy. And remembering the pleasant evening spent in Cape
Girardeau with the commandant's family, Lewis requested appointments for two sons of Louis Lorimier. The five young men from Missouri recommended by Meriwether Lewis in 1804 accounted the next
year for a third of West Point's enrollment. Four of the five graduated,
but only Charles Gratiot's son made the military his career.[26]

While in St. Louis, Lewis also met Dr. Antoine Saugrain, a Paris-
born physician who had come to the city from Louisville in 1800.
When Saugrain first moved to Kentucky, Jefferson wrote to George
Rogers Clark that the doctor had been recommended to him as "a gentleman of skill in his profession, of general science and merit."[27] Local
legends and family histories say Saugrain supplied the expedition
with a medicine chest, medicines, matches, and thermometers. An
often told story has Saugrain providing Lewis and Clark with thermometers by scraping mercury off the back of his wife's fine French
mirror, but the captains already had thermometers when they reached
St. Louis. The journals show Lewis kept temperature records on the

Ohio and Mississippi, and Clark made temperature readings part of his daily weather records at Camp Dubois. They could have used more thermometers, however, for the last one broke during the westward crossing of the Rocky Mountains in 1805, leaving Lewis to regret he was not able to keep temperature records during the winter on the Pacific Coast.

Saugrain may have supplied them instead with a few "lucifers," or strike matches. These novelties appealed to him and had already been tried by Jefferson, who found them handy for lighting a bedside candle. If they were useful at Monticello, they should have been of even more value on the expedition, but the journals make no reference to how the explorers started fires for cooking and warmth.[28]

Whatever Saugrain's tangible contributions to the expedition may have been, Lewis would have become acquainted with this distinguished member of the St. Louis community. Dr. Saugrain was a pleasant, well-educated man whose home contained an excellent library of works on botany, mineralogy, medicine, and other scientific fields, while his garden, protected by a stone wall, featured herbs, fruit trees, and two greenhouses. Both the home and its garden would have appealed to Meriwether Lewis, who was familiar with his mother's garden of medicinal herbs and familiar with Jefferson's library, which included hundreds of books on medicine and other fields of science.

Aware of Jefferson's diverse interests, the Chouteaus assembled various scientific specimens to be forwarded to the president, including silver and lead ore, which Osage war parties had brought out of Mexico, and an "elegant Specimen of Rock Chrystal, also from Mexico." Charles Gratiot contributed a horned toad, "a native of the Osage Plains," which had been found some five hundred miles southwest of St. Louis. And to this eclectic collection Dr. Saugrain added a large hairball "Taken from the Stomach of a Buffaloe." In addition to these curiosities, Lewis included for the president a map showing part of Upper Louisiana "compiled from the best information that Capt. Clark and myself could collect, from the Inhabitants of Saint Louis, haistily corrected by the information obtained from the Osage Indians lately arrived at this place." He also sent Jefferson two town plans of St. Louis and a chart of the Mississippi between St. Louis and New Orleans, all drawn by "Mr. Soulard."[29]

Antoine Soulard, a Frenchman who had served as surveyor-general of Upper Louisiana for the Spanish, was another St. Louisan who offered assistance to Lewis. He provided census figures indicating the population of Upper Louisiana was about ten thousand, of whom slightly more than five thousand were U.S. citizens who had immigrated. Somewhat more than two thousand were French or Canadian, about two thousand were slaves and other "people of colour," and only a few were Spanish. Lewis believed these figures underestimated both the total population and the number of Americans, but he knew any data of this kind would be of interest to Jefferson.[30]

Lewis and Clark were also able to obtain an English version of a map Soulard had drawn for Governor Carondelet, indicating the tribes that lived along the Missouri River and its tributaries. Historian James Ronda notes in his book *Lewis and Clark among the Indians* that Soulard's map, with its circles and triangles designating the villages and nomadic ranges of tribes along the entire length of the Missouri River, showed "with remarkable accuracy the locations of western Indians at the end of the eighteenth century." This information was always being updated as traders, boatmen, and others who traveled the Missouri between St. Louis and the Mandan villages shared what they had learned from their experiences and what they had heard about the river and the people beyond the Mandans. Ronda writes of St. Louis: "No other city could have provided Jefferson's explorers with such a range and quality of information about the Indians."[31]

James Mackay, a former explorer and fur trader living in the St. Louis area, had established a post in present-day Nebraska about eight years earlier and from there sent John Evans, another explorer and trader, farther along the river. In 1796, Evans reached the Mandan Indian villages in what is now North Dakota. Information gathered by these two men was especially valuable to Lewis and Clark. Mackay visited Camp Dubois at least once, and Lewis was able to obtain a copy of his map. He was also given Mackay's journal, written in French, and portions of it were then translated for him by John Hay, a former trader who had become the U.S. postmaster at Cahokia. Lewis and Clark carried with them on the expedition a copy of Evans's map, which even designated the places along the Missouri where bands of Sioux were most likely to be encountered.[32]

Clark spent most of the winter at Camp Dubois but was in St. Louis for several weeks in February and March. He was present at the ceremony marking the formal transfer of Upper Louisiana, and on another occasion both he and Lewis accompanied Gratiot and Pierre Chouteau a few miles up the Missouri in an effort to stop attacks on native Osage Indians by about one hundred members of the Kickapoo tribe who had crossed to the west side of the Mississippi as white advancement continued. Conflicts among tribes and between Indians and white settlers would be part of the region's history for the remainder of William Clark's life. This effort at restoring peace was a glimpse into the world he would know in the years following the expedition.

In April the captains attended a dinner hosted by Amos Stoddard, who had presided at the transfer ceremonies and had become the territory's military and civil commandant. The dinner "for about 50 Gentlemen" was followed by a ball intended to be worthy of the French style of entertaining. Clark had enjoyed a foretaste of that style during his 1797 visit to St. Louis when he was a guest at a party given by Auguste Chouteau. "I saw all the fine girls & buckish Gentlemen," Clark wrote, adding that the dance lasted until almost daybreak.[33] Six years later his prose was more subdued, but the city's love for dancing was unabated. Stoddard's party continued until nine o'clock the next morning. "No business to day," Clark observed.

Whenever both leaders were away from Camp Dubois, Sergeant John Ordway had the unenviable job of trying to control the restless and bored recruits. On learning one man had refused to obey the sergeant's commands and another had encouraged this insubordination, Lewis warned that such behavior could jeopardize "the ultimate success of the enterprise in which we are all embarked."[34] In this case, the Wood River experience was not prophetic. There were a few instances of misconduct and one desertion early in the journey, but as the party moved farther into the unknown and faced greater external challenges, camaraderie grew and disciplinary problems virtually disappeared.

On New Year's Day 1804, Clark had "put up a Dollar to be Shot for" in a test of marksmanship between his men and local residents. Although the "Countrey people" were the winners in the first contest, a winter of practice gave Clark's men the advantage. When a rematch was held in late April, he wrote: "Several Country men Came to win my mens money, in doing So lost all they had, with them." In addition to the country people and an occasional guest from Cahokia, St. Louis, or St. Charles, Indians visited Camp Dubois throughout the winter, bringing deer and other meat and sometimes remaining for the day or overnight. Clark, in turn, visited their camps, bringing them the flour, meal, and freshly caught fish they requested.

Early in 1804, Clark prepared a map of the Northwest, which left many spaces to be filled in. Jefferson had given the captains maps of the coastal area, and they had learned from their St. Louis contacts about the Missouri River as far as the Mandan villages. But neither they nor any other white men had an accurate picture of what lay between the Mandan villages and the Columbia River. Because of this, the captains seriously underestimated both the length and the difficulties of the journey, as evidenced by a schedule Clark drew up at Camp Dubois. The plan had the Corps of Discovery reaching the Rocky Mountains before the winter of 1804 and returning to St. Louis from the Northwest by December 1805.[35] In reality, almost every stage of the expedition took longer than anticipated. The explorers reached the Rockies by a route that added hundreds of miles to the initial calculations, and they returned to St. Louis nine months later than Clark had estimated.

One part of the Wood River routine was making weather observations. A weather diary begun at the start of 1804 listed not only temperatures but wind directions, river changes, and other information for each day—a collection of data that foreshadowed the explorers' remarkable documentation of their journey to the Pacific. The men had arrived in December, amid snow, hail, and violent winds. Sleet and snow fell throughout the next two months, with Clark describing one blustery January day as "truly gloomy." Blocks of ice filled the Missouri River, while ice nine inches thick closed the Mississippi.

At Wood River and during the expedition, Clark repeatedly showed an ability to look beyond personal discomforts and appreciate the world around him. In his entry for January 25, 1804, the words "I was Sick all night" are followed immediately by his observation that the branches of surrounding trees were "gilded with Ice from the frost of last night which affords one of the most magnificent appearances in nature." Weeks later he saw another of nature's magnificent and ephemeral displays, the changing colors of the aurora borealis, or the northern lights. He would witness this phenomenon again at the expedition's next winter camp, at Fort Mandan. The sentry who awakened him must have known William Clark would readily stand in the cold to watch what his journal describes as floating columns of brilliant light in the night sky.

Weather records Clark kept at Wood River show his search for harbingers of spring. By February the geese and ducks of various kinds had returned. Warm, fair weather in March brought the sound of frogs and the sighting of the first white cranes. By early April the spicewood and violet had appeared, and summer ducks were arriving. On April 17, Clark wrote: "The trees of the forest particularly the Cotton wood

The Mississippi River as seen from the Lewis and Clark State Historic Site south of Hartford, Illinois. The confluence of the Mississippi and Missouri Rivers is now about four miles south of its 1804 location. *Ann Rogers.*

begin to obtain . . . a Greenish Cast at a distance." The long winter at Camp Dubois had ended.

The pace of preparations quickened with the coming of spring. The captains issued orders naming the men chosen for the permanent party and dividing them into three squads to serve under Charles Floyd, Nathaniel Pryor, and John Ordway. Pryor and Floyd, both recruited by Clark in Kentucky, were made sergeants, a rank already held by John Ordway, who had joined from Captain Russell Bissell's company of the First Infantry at Kaskaskia.

A letter Sergeant Ordway wrote on April 8, 1804, to his parents in New Hampshire makes a good preface to the journal he kept during the expedition. After assuring them he was healthy and "in high Spirits," he told them he was on a journey with "Capt. Lewis and Capt. Clark," who had been appointed by the president to "ascend the Missouri River with a boat as far as it is navigable and then to go by land," if nothing prevented them, "to the western ocean." He explained that the captains had chosen for this expedition men from both military and civilian life, adding proudly: "I am So happy as to be one of them pick'd Men from the armey." Then, considering the possibility that he might not return, he let his family know how they could collect any pay due him by the United States, as well as two hundred dollars in cash he had left for safekeeping at Kaskaskia. The letter closes with his promise to "write next winter if I have a chance."[36]

In the final weeks at Camp Dubois, supplies were gathered and packed. The hunters would be shooting game as they moved upriver, but the Corps could not expect to live entirely off the land. Clark's shopping list included flour and salted pork, along with biscuits, parched corn, beans, salt, sugar, coffee, and whiskey. Lewis had brought some dried soup from Philadelphia, but most of the food assembled at Wood River came from local suppliers on both sides of the Mississippi.

John Hay, the trader who had earlier assisted as an interpreter in Lewis's meetings with the Spanish authorities and later translated some of Mackay's journal, arrived from Cahokia to help Clark assemble, bundle, and store the flags, medals, and assorted presents to be distributed to tribes along the Missouri and beyond. Numerous and diverse as the gifts were, Clark worried that they were "not as much as I think necssy for the multitud of Inds. tho which we must pass on our road across the Continent."

As he moved ahead with the preparations, Clark received disappointing news from Meriwether Lewis. "My dear friend," the letter of May 6 begins, "I send you herewith inclosed your commission accompanyed by the Secretary of War's letter; it is not such as I wished, or had reason to expect." Both men had anticipated the War Department would assign Clark the rank of captain; instead, he was to be a lieutenant. The news was unsettling to Lewis, who had invited Clark to participate with the assurance the president would "grant you a Captain's commission which of course will intitle you to the pay and emoluments attached to that office." In his May 6 letter, an embarrassed Lewis could offer Clark some solace in that regard: "You will observe that the grade has no effect upon your compensation, which by G__d, shall be equal to my own."[37]

The remaining question was how Clark would be addressed on the expedition. He had been "Captain Clark" in Lewis's journals since the keelboat left Louisville and "Captain Clark" to the men he was training at Camp Dubois. Lewis, who had given him every reason to assume they would be sharing command on the journey to the Pacific, now offered a solution: "I think it will be best to let none of our party or any other persons know any thing about the grade."[38] Throughout the expedition he would refer to his coleader as "Captain Clark," and there is no evidence that any of their men knew otherwise.

Whatever resentment Clark felt about the rank (and he later acknowledged he felt some), it did not slow his activities. The day after receiving Lewis's letter, he practiced taking compass bearings and tested the keelboat on the Mississippi while twenty oarsmen rowed the boat several miles to the north. Because the test run revealed problems with balance, the next few days were given over largely to rearranging cargo, hard work made more difficult by hot weather. Clark's response was to have drinking water brought over from the west side of the river, for at Camp Dubois, just below the confluence, the "much Cooler" waters of the Missouri River had not yet blended with those of the Mississippi.

The keelboat's appearance had changed since Clark sketched it on his maps of the Mississippi's Grand Tower. The detailed drawings he made at Wood River are bordered by his notes on modifications, including the addition of lockers to provide safe storage for cargo. Because the tops could be raised, they would also provide a measure of defense if the boat were attacked. There were two swivel guns mounted on the keelboat's stern, another on the six-oared white pirogue, and another on the seven-oared red pirogue. Mounted at the bow of the keelboat was a small cannon that Lewis apparently purchased in the St. Louis area.

Some of the men recruited from the military carried army muskets, some of Clark's recruits brought Kentucky long rifles, and some of the men would use the Model 1803 rifles Lewis had purchased at Harpers Ferry. Clark ordered "every man to have 100 Balls for ther Rifles & 2 lb. of Buck Shot for those with mussquets."

On May 11, George Drouillard arrived at the camp with seven French *engagés,* experienced boatmen hired by the Chouteaus to travel with the Corps of Discovery as far as the Mandan villages. Like the seven men Clark recruited from the Louisville area, the engagés were young, hardy, and free to embark on a long expedition. Unlike Clark's seven young men, they knew both the Missouri River and the Indian tribes Lewis and Clark would encounter in the early portion of their journey. In fact, many engagés were related by blood or marriage to Indian women. These seven were assigned to crew the larger of the pirogues.

Two days later Clark reported in a letter to Lewis that the boats were in order, the provisions stored, and the men "all in health and readiness." After more than five months of preparing and waiting, the Corps of Discovery would at last set out from Camp Dubois.

# 2

# WESTWARD ACROSS MISSOURI

O n the afternoon of May 14, 1804, the keelboat and two pirogues crossed the Mississippi to the mouth of the Missouri and headed upriver. No dignitaries were present for the departure, simply the people who had been the Corps of Discovery's neighbors during the winter at Camp Dubois. Perhaps the woman who laundered their clothes was there and the man whose wagon hauled logs to build the fort. Clark had decided to take the boats as far as the village of St. Charles, some twenty miles to the west, and wait for Lewis to complete his preparations in St. Louis and come by land to meet them. This plan would allow Clark time to make any needed changes in the loading of cargo and attend to other final details while awaiting Lewis's arrival. Because the boats did not set out until four o'clock in the afternoon, they traveled only about four miles before the men made camp opposite a small creek referred to in Clark's journal as "Cold water."

Rain had fallen during the day and continued through the night, extinguishing their campfires and soaking some items stored in the pirogues. Soon after the flotilla set out the next morning, it became clear one of the pirogues lacked the manpower to keep up. There were also problems with the keelboat, which would strike submerged logs three times as it made its way against the fast current of the Missouri. The boat was too heavily loaded in the stern and was riding up on the logs at the risk of damaging its hull. Clark realized that cargo would have to be rearranged to bring the bow lower in the water.

On the third day of the shakedown cruise, the boats reached St. Charles, where "a number [of] Spectators french & Indians flocked to the bank to See the party." Sergeant Ordway described the town as "an old French Settlement & Roman Catholick," with "Some Americans Settled in the country around." The villagers, who numbered about 450, impressed Clark as "polite & harmonious." Most lived in simple houses on a street parallel to the river. Their small gardens were generally tended by old men or young boys, since most men in the prime of life were hunters or hired boatmen, away from the village for long periods.

While waiting for Lewis, Clark enjoyed local hospitality at the home of François Duquette, "an agreeable man" with "a Charming wife." Although he had suffered financial reverses as a merchant, Duquette lived in what his guest found "an eligent Situation on the hill Serounded by orchards" and an excellent garden. A dozen years after Clark's visit, the home's setting would be described in greater detail: "The town is partly visible from this retirement. . . . The river spreads out below it in a wide and beautiful bay. . . . The trees about the house were literally bending under their load of apples, pears, and yellow Osage plums. Above the house and on the summit of the bluff is a fine tract of high and level plain covered with hazel bushes and . . . a great abundance of grapes."[1]

Clark's five days in St. Charles gave him little time to savor the tranquil surroundings. The keelboat became a hub of activity, as some people came just to visit, while others brought fresh vegetables from their gardens. The French boatmen arrived with eggs and milk, items that would soon disappear from their diet, and Duquette sent a gift of fish. Clark spent much of his time supervising the reloading of both the keelboat and one of the pirogues to make them better able to handle the hazards the river would present.

Another step in lessening the dangers of the voyage was the enlistment at St. Charles on May 16 of Pierre Cruzatte. More than once this able and experienced riverman would save the expedition's boats from disaster. Unlike the engagés, he was assigned to the keelboat on the first portion of the journey and would remain with the expedition on the trip to the Pacific.

Clark, as he would often do in the miles ahead, measured the width of the Missouri, which he found to be 720 yards across at St. Charles, and he used a sextant in an effort to determine the latitude. After writing a letter to Lewis and dispatching this with Drouillard, he made a record of the day's events, a practice he would continue throughout the journey.

One incident he no doubt regretted having to report involved misconduct by three of his party. Anticipating possible trouble, he had cautioned the men to have "a true respect for their own Dignity and not make it necessary . . . to leave St. Charles . . . for a more retired Situation." But after their months of confinement at Camp Dubois, some found the town's temptations more than they could handle. William Werner and Hugh Hall were charged with being absent without leave, and John Collins not only took unauthorized leave but also was accused of behaving in an "unbecomming manner" at a dance and then speaking disrespectfully on his return. A court-martial was convened in which Sergeant Ordway and four privates found the men guilty of the charges but recommended leniency for Werner and Hall. Collins, who had already caused problems at Wood River, received a sentence of fifty lashes, and the three were returned to duty.

On Sunday morning, about twenty members of the expedition attended a mass at the log church of St. Charles Borromeo. Beneath its floor were the graves of the village's founder, Louis Blanchette, and his wife, a member of a Missouri River tribe. No one is sure when Blanchette made his earliest visit to the area, but apparently by 1770 the French Canadian trader, along with a group of relatives and friends, established the first white settlement on the north bank of the Missouri. He called the place "Les Petites Côtes," or "the Little Hills." When the region came under the control of the Spanish, the village and its church were renamed for the patron saint of Spain's King Carlos. In writing about the town, Clark used both the name St. Charles and the more descriptive name given by Blanchette.

Captain Lewis spent that Sunday traveling overland to St. Charles in the company of a send-off delegation of prominent St. Louisans, including Auguste Chouteau, Amos Stoddard, Charles Gratiot, and Dr. Antoine Saugrain. The weather had been pleasant when Lewis said good-bye to other friends, thanked "that excellent woman the spouse of Mr. Peter Chouteau" for her hospitality, and left the Chouteau home. The first five miles of the trip from St. Louis took him through land he described as "a beatifull high leavel and fertile prarie which incircles the town . . . from N.W. to S.E." Then, less than two hours into their journey, a violent afternoon thunderstorm forced the travelers to take shelter in a cabin. Ninety minutes passed and the skies gave no sign of clearing. Lewis, determined to reach St. Charles by nightfall, set out in the rain with most of the party still accompanying him. Within a few hours he had joined Clark and learned their men were "in good health and sperits." The captains dined that evening with Charles Tayon, the former commandant at St. Charles, but Lewis excused himself early to rest aboard the keelboat and prepare for departure the next day.

Both leaders were present, their men were assembled, the cargo had been properly reloaded, and additional supplies had been purchased and stored. Because the Louisiana Territory now belonged to the United States, the party could be much larger than originally conceived. There

The home built by Daniel Boone's youngest son, Nathan, where the elder Boone died in 1820, is near Defiance, Missouri. *Ann Rogers.*

were the "nine young men from Kentucky"—the woodsmen Clark recruited and two men who had come down the Ohio with Lewis. There were soldiers recruited over the winter from various posts. The Chouteaus had engaged seven French boatmen, and at least two men were enlisted at St. Charles. York would function as a member of the crew, and Drouillard would serve as a hunter and interpreter. Altogether about forty-seven men were now ready. On the afternoon of May 21, as those gathered on the bank cheered, the Corps of Discovery embarked on the Missouri River.

At their campsite the next evening, they were joined by a group of Kickapoo Indians who lived on the Illinois side of the Mississippi but had been in St. Charles while the Corps of Discovery was there. Clark wrote that they "told me Several days ago that they would Come on & hunt and by the time I got to their Camp they would have Some Provisions for us." This prearranged meeting, in which the captains obtained four deer in exchange for two quarts of whiskey, proved to be almost the only contact with Indians the expedition had during the westward crossing of Missouri. A week later one of the hunters reported seeing six Indians while he was away from the main party, and there would be several more accompanying boats the explorers met on the river. But with these exceptions, the next encounter with Indians would take place more than ten weeks and six hundred miles later, in what is now Nebraska.

On May 23 the boats reached the Femme Osage River, which Clark translated as the "Woman of Osage River" and the "Osage Womans" River. A stop was made to pick up Reubin and Joseph Field, who had been sent ahead by the captains to purchase butter, corn, and other supplies. Thirty to forty families were living along this stream, and Clark wrote that "many people Came to See us." Although Lewis noted that "this part of the country is generally called Boon's settlement," the expedition's journalists made no mention of seeing its most famous resident.[2] When Daniel Boone came to the Femme Osage region from Kentucky five years before the expedition passed through, Spanish officials gave him a land grant of a thousand arpents, equal to about 850 acres, with additional land grants for family members and others who followed him. Soon the government appointed him commandant of the Femme Osage region, a position of civil and military authority not yet affected by the Louisiana Purchase.

Many years before, Boone had mastered the skills Lewis and Clark valued and sought in the men they recruited. An excellent woodsman and hunter, he knew Indian ways and had an instinctive sense as a pathfinder. In 1775, while Meriwether Lewis was in his first year of life, Boone blazed a 250-mile trail through the Cumberland Gap to the Kentucky River, where he established the settlement of Boonesborough. The Wilderness Road, as the trail was called, became one of the leading pioneer routes. In 1804, he was seventy years old, the elder statesman of his Missouri River community. But if the inveterate traveler were among those who watched as the Corps of Discovery moved upstream, he almost certainly would have been stirred by a longing to join the explorers on their epic journey.

That journey came close to a tragic and premature end only a few hours later near the present site of St. Albans in Franklin County. Limestone bluffs rose high above the south bank of the Missouri, and at the base of the bluffs was a cave known to rivermen as the Tavern. In a space Clark described as "about 120 feet wide 40 feet Deep & 20 feet high," French traders and other travelers would take refuge from heat, cold, or storms. He noted that "many nams are wrote on the rock" and

Late afternoon sun spotlights the bluff where Lewis narrowly escaped a deadly fall. In 1804, the river flowed directly below the bluff at St. Albans. *Ann Rogers.*

"many different immages are Painted on the Rock at this place" where both "the Inds & French pay omage."[3]

Lewis, as part of his exploration, climbed onto a jagged outcropping of rock at the top of the bluffs. Three hundred feet above the swift current of the Missouri River, which in 1804 flowed directly below, he suddenly lost his footing. Meriwether Lewis's death, or even serious injury, at this early stage of the journey would almost certainly have marked the end of the expedition. Yet, in the terse style that is a trademark of his journals, Clark wrote only that he "Saved himself by the assistance of his Knife" and "caught at 20 foot." Just over a year later, in present-day Montana, Lewis described a similar mishap: "In passing along the face of one of these bluffs today I sliped at a narrow pass . . . and but for a quick and fortunate recovery by means of my espontoon I should been precipitated into the river down a craggy pricipice of about ninety feet."[4] The captains may have decided not to mention the Tavern Rock incident to the rest of their party. None of the other journal keepers recorded it.

More perils awaited them on the water. Melting snow upriver combined with spring rains to make the Missouri high and its current unusually fast. Banks undermined by the current's force frequently caved in, sending trees and large sections of earth into a boat's path. Many trees uprooted this way became embedded beneath the murky surface of the water. Known to rivermen as "sawyers," they were capable of ripping open a vessel's hull. Sandbars were also common hazards, lying partially obscured and shifting with the current. The crews tried to keep their boats away from the main channel where the current was strongest, but close to the shore, there were overhanging branches that could snare or break a mast.

An early test of the boatmen's skills came a day after passing Tavern Cave. Just beyond a place known as the Devil's Raceground, where "the Current Sets against Some projecting rocks for half a mile," the keelboat was forced to veer away from crumbling banks on the south side of the river. While trying to pass between an island and the north bank, the boat was driven onto a sandbar by the swift current, which turned the vessel and broke its towrope. Clark's journal details the crew's efforts to save the craft as it threatened to capsize: "All hand Jumped out on the upper Side and bore on that Side untill the Sand washed from under the boat and . . . by the time She wheeled a 3rd Time got a rope fast to her Stern." Crewmen were then able to pull the boat into navigable water. After "So nearly being lost," Clark called that stretch of river "the worst I ever Saw."

On May 25, having traveled about fifty miles beyond St. Charles, the expedition arrived at the French settlement of La Charette. The site, on the north bank of the river near present-day Marthasville, has since been obliterated by flooding and river changes. Some of the expedition's journalists referred to it as St. John's, a name derived from the Spanish fort of San Juan del Misuri, which had stood in the area a few years earlier.[5] In the spring of 1804, La Charette consisted of about seven small cabins, whose occupants lived primarily by hunting. Its significance to Lewis and Clark lay in the fact that it was "the Last Settlement of Whites" on the Missouri River. More than two years would pass before they would again see this or any other white community.

While at La Charette, the captains had an opportunity to speak with Régis Loisel, a French Canadian trader who had come to St. Louis about ten years earlier. When he met the expedition, Loisel was returning from his post at Cedar Island, twelve hundred miles upriver in present-day South Dakota. After their talk, Clark wrote that he had

The Missouri River, shown here about sixty miles west of the confluence with the Mississippi, was the expedition's principal highway on its journey to the Pacific. *Ann Rogers.*

given them a substantial amount of information, telling them, among other things, that he had seen no Indians since passing a Ponca village, well above what is now Sioux City, Iowa.

The day after leaving La Charette, the boats met a pair of canoes loaded with pelts and buffalo robes coming down from the Omaha nation, 750 miles upriver. That same morning they met another pair of boats, also loaded with furs and pelts, one coming down from the Pawnee, another from the Grand Osage. Apparently little was learned from these traders, but a meeting four days later with a boat coming from the Osage brought disturbing news. The French trader camped for the night with the Corps of Discovery and told Lewis and Clark that some Osages had burned a letter sent by the Chouteaus informing them of the territory's transfer to the United States. The captains could only wonder what reception they would receive when they finally met Indians in tribal force.

Whatever awaited them on land or river, the men would be functioning within a framework of military discipline. As part of that framework, Captain Lewis, on May 26, issued Detachment Orders that defined the responsibilities of the expedition's members, especially the three sergeants: John Ordway, Charles Floyd, and Nathaniel Pryor. When the keelboat was under way, the orders stated, the sergeant at the helm would steer the boat, attend the compass when necessary, and make certain that the deck was in order and all baggage properly stored. The sergeant at the center would, among other assignments, "manage the sails, see that the men at the oars do their duty," and "keep a good lookout for the mouths of all rivers, creeks, Islands and other remarkable places." The sergeant at the bow was assigned to "keep a good look out for all danger," reporting obstructions in the river, other boats, and any evidence of Indians. Using a setting pole, he would also "assist the bowsman in poling and managing the bow of the boat." The sergeants, who rotated their positions in the boat every morning, were "relieved and exempt from all labour of making fires, pitching tents or cooking" but were to "direct and make the men of their several messes perform an equal propotion of those duties." Cooking was done only in the evening, and no pork was to be issued if fresh meat was available. John Ordway was in charge of issuing the provisions and making arrangements for guard duty.

In addition, the sergeants were "directed each to keep a seperate journal . . . of all passing occurences" and "observations on the country . . . as shall appear to them worthy of notice." Jefferson wanted the information gathered on the voyage kept in several copies as protection against "accidental losses," and having the sergeants keep journals was one way Lewis found to provide this protection. Some other members of the expedition took up the practice, and during the westward crossing of Missouri at least seven men were writing.[6]

May 27, the day after Lewis issued his Detachment Orders, the expedition arrived at the mouth of the Gasconade River, which Clark described as "a butifull stream of clear water." While the expedition camped on a willow island in the mouth of the river, he measured and found the Gasconade to be 157 yards wide and nineteen feet deep near its entrance to the Missouri. In Lewis's "Summary View of Rivers and Creeks," a report on the tributaries of the Missouri, he wrote that the river "is much narrower a little distance up, and is not navigable, (hence the name gasconade)."[7] He had no doubt read or been told that the French name, which implies bragging or boasting, was bestowed because the river did not measure up to its initial promise. Some believe Frenchmen in the region named the river for the Gascony province of

The Gasconade River was described by Clark as "a buti-full stream of clear water."

*Ann Rogers.*

southwestern France, whose residents were reputed to be braggarts.[8] Lewis, again drawing on his sources, added that "the country watered by this river, is generally broken, thickly covered with timber and tolerably fertile." Sergeant Ordway gave a more positive report in his journal: "A handsome place, the Soil is good, the Country pleasant."

During the two-day stay, Lewis tried to complete celestial observations delayed by heavy clouds on his first attempt; and the engagés dried and restored to the red pirogue the clothing and provisions drenched, seemingly through their carelessness, during a night's rain. Arms and ammunition were inspected, and several men took advantage of the well-timbered confluence area to hunt. George Shannon and Reubin Field each killed a deer, while "one Man fell in with Six Indians hunting." When the Corps of Discovery was ready to leave, late in the day on the twenty-ninth, one hunter still had not returned.

The engagés and their red pirogue were left to wait for the missing man, while the other boat crews continued up the Missouri for a few miles. During the night they could hear the sounds of gunfire as the Frenchmen tried to help the apparently disoriented hunter find his way back to the river. Ordway's journal identifies the man as Joseph Whitehouse, but it is Whitehouse's own journal that solves the mystery of his whereabouts: "As I was a hunting this day I came across a cave on the South Side . . . about 100 yards from the River. I went a 100 yards under ground. had no light in my hand if I had, I Should have gone further their was a Small Spring in it. it is the most remarkable cave I ever Saw, in my travels."

While the red pirogue waited for Whitehouse, the keelboat and white pirogue passed a cave on the north side of the Missouri known as Montbrun's Tavern, which, like Tavern Cave at St. Albans, was used as a refuge by river voyagers. The expedition's journals mention only a few of the state's thousands of caves, formed by acidic water seeping through the region's porous limestone. Their number has given Missouri the nickname "the Cave State."

The Corps of Discovery spent the first three days of June at the mouth of the Osage River, named for the Indians who in 1804 lived near its headwaters and had earlier lived near its confluence with the Missouri. The captains carried a map obtained from Antoine Soulard that showed the entire Osage region, and they also would have learned about the confluence area from the Chouteaus, who operated a trading post from 1794 to 1802 on the south bank near some Osage villages.[9]

In preparation for taking celestial observations at the confluence, Lewis immediately set a detail of men to the task of felling trees on the point of land between the two rivers. He and Clark worked past midnight and throughout a good part of the following two days, but clouds obscured the moon by night and the sun by day. Lewis's list of figures is interspersed with notations such as "probably a little inaccurate" and "the observation completely lost." Despite the difficulties and frustrations, Lewis knew Jefferson's instructions included the directive to "take observations of latitude & longitude, at all remarkeable points on the river, & especially at the mouths of rivers,"[10] and so the ritual of taking celestial observations would be repeated at major confluences.

For the document he was preparing similar to Lewis's "Summary View," Clark measured the width of the two rivers, finding the Osage to be 397 yards wide and the Missouri 875 yards wide. He also enjoyed a "delightfull prospect" from atop a hundred-foot hill, which he described as being about 80 poles (440 yards) from the point and having "two Mouns or graves" on "high land under which is a limestone rock." A hill matching Clark's description in all details is at the entrance to Smoky Waters Conservation Area, near Osage City.[11]

The three-day stay at the Osage was also a time for reassembling. The red pirogue arrived, the Frenchmen having retrieved Joseph Whitehouse, and, at about the same time, two other men arrived exhausted from a week of hunting. Clark's journal entry for the second day of June recounts their experience: "George Drewyer & John Shields who we had Sent with the horses by Land on the N Side joined us this evening much worsted, they being absent Seven Days depending on their gun, the greater part of the time rain, they were obliged to raft or Swim many Creeks." In spite of their ordeal, Clark was able to add that the hunters "gave a flattering account of the Countrey" they had seen.

Their tribute accords with the observations of those who kept journals during this part of the expedition, and of these none is more consistent in his praise of the Missouri landscape than Sergeant Charles Floyd, a Kentuckian. "The Land is Good," Floyd wrote, as the boats approached the Osage confluence, and the phrase became a theme for his descriptions. He called the area near present-day Jefferson City as "Butifull a peas of Land as ever I saw" and wrote as they continued through central Missouri: "This is a butifull Contry of Land." A week later, in what is today Saline County, he admired as "handsom a prarie as ever eney man saw." In the weeks ahead, his journal entries would repeat the theme: "The Land is Good."

The verdure of the landscape was due in part to drenching spring rains that had fallen at least seven of the fourteen days since the Corps of Discovery left St. Charles. Captain Clark reported he was suffering from "a bad Cold with a Sore throat," but typically it did not reduce his activities or lessen his interest in the country he was seeing. On the afternoon of June 4, when the boats had traveled about twenty-two miles beyond the Osage, he left his usual station in the keelboat and walked for more than a mile through rushes and chest-high nettles along the south bank. He had been told by the Frenchmen in the party that a nearby hill contained lead ore, but his own search revealed no sign of the mineral. Lewis's decision to make camp just below gave Clark more time to explore. Halfway up the 170-foot hill, under a limestone outcropping, he found a "verry extensive Cave," and at the summit he discovered an Indian burial mound about six feet high. By scaling to a spur of rock overhanging the water, he had a view of the Missouri for twenty to thirty miles upstream.[12]

From the river the next day the expedition saw an Indian pictograph on a rock. Clark made a small sketch of the figure, which appeared to be a "Manitou," or spirit, with the head and shoulders of a man and the antlers of a deer. Sergeant Ordway called it a "pickture of the Devil." Ordway may have felt bedeviled. He had been at the helm the day before when the keelboat's mast broke, making it impossible for the crew to hoist the sail and take advantage of a favorable wind. The mishap is candidly explained in his journal: "Our mast broke by my Stearing the Boat near the Shore. the Rope or Stay to the mast got fast in a limb of a Secamore tree & it broke verry Easy." The expedition's other writers recorded the incident in an impersonal way, noting simply that "our mast broke" or that the "Stersman Let the Boat Run under a lim." While Clark's account is more specific in its mention of "the Sergt. at the helm," even he doesn't identify the sergeant. Ordway's journal is the only one to name him as the person responsible for the accident.

John Ordway, always willing to shoulder responsibilities, may have been too quick to blame himself. The Detachment Orders issued by Captain Lewis on May 26 stated it was the duty of the sergeant at the bow "to keep a good look out for all . . . obstructions" and "notify the Sergt. at the helm." Typically, Clark did not seem to dwell on the misfortune. The notation "broke the mast" is followed immediately in his journal by the phrase "Some delightfull Land." But the mishap did lead Clark to give the name Mast Creek to a stream believed to be present-day Grays Creek, flowing into the Missouri at Jefferson City.

Another stream named on June 4 commemorated a happier event. "We named Nightingale Creek," Clark wrote, "from a Bird of that discription which Sang for us all last night, and is the first of the Kind I ever heard." Because there was no authentic nightingale in America, Lewis and Clark scholars have been left wondering and debating just what bird the expedition heard sing. The candidates include, among others, the mockingbird, catbird, hermit thrush, whippoorwill, chuck-will's-widow, and cardinal. The last of these was also known as the Virginia nightingale, but its song would have been familiar to the Virginia-born captains. The case made for the whippoorwill finds support in a passage written by a nineteenth-century English visitor to America: "If the name nightingale were to be given to any of the feathered race in the southern states, that called the `Whip-poor-Will,' is best entitled to it. This bird sings a plaintive note almost the whole night long."[13]

York is one of the figures portrayed in Eugene Daub's statue on Quality Hill in Kansas City. *Ann Rogers.*

After passing Little Moniteau Creek, named for the figure on the rock, the boats came to a sandbar several miles in length. Clark wrote that "York Swam to the Sand bar to geather greens for our Dinner and returnd with a Suffcent quantity" of wild cresses and tongue grass. Although the hunters had brought in seven deer the previous day and the venison had been jerked before the boats set out in the morning, the captains recognized the need to eat greens whenever possible to augment a diet heavy with meat. At the Gasconade River, Lewis had collected "a species of cress which grows very abundantly alonge the river beach in many places," noting that "my men make use of it and find it a very pleasant wholsome sallad."[14]

The sandbar that grew their dinner greens also forced the keelboat to drop back two miles when rapidly shifting sands made passing on the left side of it too difficult. Clark briefly lamented that they "had a

The expedition passed the Big Moniteau Bluffs on June 7, 1804, where they saw the Indian pictographs Clark described and sketched. *James Denny.*

fine wind, but could not make use of it, our Mast being broke." Without it, they still made twelve miles. While they waited at their campsite for one of the pirogues, which needed two hours to catch up, a man sent to scout the area reported he had found fresh signs of about ten Indians. Clark speculated they were probably Sauks on their way to war against the Osage nation.

After the mast was repaired, the boats continued to Salt Creek, which Lewis noted in his "Summary View of Rivers" was navigable for pirogues for forty to fifty miles, passing through "delightfull country intersperced with praries." He had learned in St. Louis or from the French boatmen that "so great is the quantity of salt licks and springs on this river," including one lick and spring about nine miles from the Missouri, that the waters of Salt Creek "are said to be brackish at certain seasons of the year." After traveling fourteen miles, the main party camped earlier than usual to allow one of the pirogues to catch up.

Along with his field notes, Clark sketched an Indian pictograph showing a man, a buffalo, and a Manitou. Natural forces and railroad blasting have destroyed the pictographs that he and other nineteenth-century travelers described. *Yale Collection of Western Americana, Beinecke Rare Book and Manuscript Library.*

On June 7, near the mouth of the Big Moniteau, the men came on a group of pictographs Clark described as "Courious Paintings and Carveing in the projecting rock of Limestone inlade with white red & blue flint of a verry good quallity." Earlier in the day they had seen evidence of buffalo in the area, and Clark now copied into his field notes one drawing that showed a buffalo flanked by an Indian and a Manitou. Scrutiny of the cliff yielded more than expected. "We landed at this Inscription," Clark wrote, "and found it a Den of rattle Snakes, we had not landed 3 minutes before three verry large Snakes wer observed on the Crevises of the rocks & Killed." When the hunters returned that evening with three black bears, the first shot by the expedition, there was a full day's ration of zoological information to record.

Setting out at dawn from their camp at Bonne Femme Creek, the men passed a number of deer licks before coming to the Lamine River, near present-day Boonville. Clark used the English translations, calling them "Good Womans River" and the "Mine River," but both are better known today by their French equivalents. The latter had been named in the eighteenth century by Frenchmen who were told Indians mined lead along its banks.[15] The captains explored the region on foot, with Lewis inspecting the area just above the Lamine, while Clark and Sergeant Floyd made a four-mile survey of the area below the Lamine's junction with the Missouri. Clark wrote that he "found the land Verry good . . . and Sufficiently watered with Small Streams." Lewis's notes were used for his later report on the tributaries of the Missouri: "The courant of this river [Lamine] is even and gentle. The country through which it passes is generally fertile, and consists of open plains and praries intersperced with groves of timber. near it's entrance, the country is well timbered and watered, and the lands are of a superior quality."

The following day, June 9, the expedition reached a prairie known then as the Prairie of Arrows. The name Pierre à Flèche appears on French maps as early as 1723, and while legends abound concerning the origin of the name, the most likely explanation and one supported by archaeological evidence is that Indians used flint obtained here to fashion arrowheads and spears. The area was destined to become an important one in the history of nineteenth-century Missouri. Clark saw it as a good location for a fort when he revisited in 1808, and a dangerously exposed post farther to the west was temporarily moved to this site in 1813. The town of Arrow Rock was the western terminus of the Boonslick Road, which linked the Boone brothers' saltmaking operation to St. Charles, and the town became the central Missouri link to the Santa Fe Trail.[16]

The passage through this area was marked by navigational problems. The keelboat caught on a snag soon after setting out in the morning, and later that afternoon another snag posed a greater hazard. The "Stern Struck a log under Water & She Swung round on the Snag, with her broad Side to the Current," Clark wrote, causing "a disagreeable and Dangerous Situation, particularly as immense large trees were Drifting down and we lay imediately in their Course." Some of the men were quickly in the water, swimming ashore with a towrope, and within a few minutes they were able to pull the boat free. The crew had responded well, and Clark now believed he could say "with Confidence that our party is not inferior to any that was ever on the waters of the Missoppie." But problems continued when the crews made camp for the night. Their hunters were left stranded on the opposite side because the banks were too unstable for a pirogue to make a safe approach and pick up the men.

The next morning, with the riverbanks still collapsing and sending

The expedition's journals during the westward crossing of Missouri are filled with descriptions of the prairies and repetitions of the phrase "the land is good." *Casey Galvin.*

large trees into the water, the expedition's journalists turned to some pleasant scenes. "A number of goslings" were sighted in the morning, and, just below the Chariton River, grapevines and "plenty of Mulbery Trees" with ripe berries caught the attention of Sergeant Ordway. A few miles above the Chariton, in present Saline County, the captains walked for about three miles through rich, open prairie, where wild plums three times the size of other plums hung "in great quantities on the bushes." These were the Osage plum, which Clark was told were "finely flavoured." He had lived since boyhood in Kentucky, where prairies called "barrens" were, in his words, "Void of every thing except grass." Looking on the fruitfulness of this Missouri prairie, he expressed surprise at the "good Land and plenty of water."

The wind blew so fiercely from the west the following day that travel, already difficult against the current, was impossible. While they waited, the men cleaned their weapons, checked on provisions, and put water-soaked articles out to dry. The hunters went in search of game, and the surplus meat was jerked. As Clark explained in his journal, it was their "Constant Practice to have all the fresh meat not used, Dried in this way." In the first eleven days of June, at least twenty-three deer and five bears had been brought in, and while the successful hunters are not always identified in the journals, the names of George Drouillard (usually spelled "Drewyer" or some variant of that) and Reubin Field appear most often. Whitehouse had written on June 8 that Drouillard killed five deer before noon.

On June 11, Ordway recorded that "Drewyer & Several more went out in the Praries a hunting, & Drewyer killed two Bear & one Deer. R Fields killed one Deer." In fact, Drouillard shot all five of the bears killed during this period. Bears were valued more often as a source of needed grease than of meat, but Clark's notation that they "had the meat Jurked and also the Venison" suggests that in this instance they intended to eat the bear meat.

They were in a land of plenty, enjoying a day's respite from their struggle against the river. Evening found the men "verry lively Danceing & Singing." No doubt they danced to tunes played by Pierre Cruzatte, who was not only a fine boatman but also a good fiddler. On numerous occasions throughout the journey, his music would enliven the celebrations of holidays or lift the party's spirits during long winter encampments and other periods of stress. To the Indians they met along the route, watching the men dance to Cruzatte's fiddle was entertainment and a glimpse into another culture.

When the boats were able to continue upstream the next day, the expedition encountered traders returning from trading with the Sioux, their boats loaded with furs and buffalo grease. In the week just past, the Corps of Discovery had met other trading parties, including Frenchmen coming down from the Kansas River and three more men returning with furs collected along the Big Sioux River to the north. From the group met on this day, Lewis purchased moccasins and three hundred pounds of buffalo grease, grease that could be used for cooking, mosquito repellent, and even boat caulking. But of greater interest to the captains was a man aboard one of the canoes who knew the language of the Sioux and was trusted by them. In Pierre Dorion, Lewis and Clark saw their best hope of persuading some of the chiefs to visit President Jefferson. Apparently without much hesitation, Old Dorion, as he was called, agreed to reverse course and travel with the expedition as far as the Sioux nation.

On June 13 the Corps of Discovery passed a bend in the river and came on an extensive prairie that was the ancient home of the Missouri Indians, a people whose tragic history was known to Lewis and Clark. The Missouri had been the predominant tribe in the region until attacks during the 1790s by Sauk and Fox war parties inflicted heavy losses. Another deadly enemy, smallpox, completed the annihilation. A few survivors fled their homeland to join other tribes, and the once-proud Missouri ceased to exist as an independent nation. The expedition camped at the mouth of the Grand River, near the present town of Brunswick. From atop a hill near their campsite the captains enjoyed "a butifull prospect of Serounding Countrey," and Lewis continued his efforts to determine the latitude and longitude of important sites.

Just beyond the Grand River, Lewis found the landscape "a happy mixture of praries and groves, exhibiting one of the most beatifull and picteresk seens that I ever beheld." He noted that "old Fort Orleans is said to have stood on . . . an Island a few miles below this place" but added that "no traces of that work are to be seen." The fort had been built by the French to keep control of the Missouri River and discourage Spanish incursions. Its location near the villages of the Little Osage and Missouri was intended to strengthen relations with those Indians and win them as allies against the Spanish. Erected in 1723, the post, which included officers' quarters, a chapel, a store, a powder magazine, and a few houses, was constructed forty years before Pierre Laclède chose the site for St. Louis.[17] Fort Orleans was abandoned after only five years, and river changes soon removed all traces of its presence; today, this very early European establishment in Missouri is commemorated by a marker near the town of Brunswick.

The expedition made slow progress along this stretch of the river. Just west of the Grand River, the boats encountered the most treacherous sandbars the men had seen so far. The swift current and collapsing banks forced the crews to thread a perilous course between two moving islands of sand. Despite their best efforts, the keelboat struck the point of one and was nearly overturned. The next day the boat twice escaped disaster when it first wheeled on a sawyer, and then proceeded on, with the river rising and the current increasing, to more sandbars between two islands. Calling it "a verry bad place," Clark wrote: "We were nearly being Swallowed up by the roleing Sands over which the Current was So Strong that we Could not Stem it with our Sales under a Stiff breese in addition to our ores." With both sails and oars insufficient, the men resorted to pulling the boat with the towrope.

The river had taken its toll, and Clark began scouting the banks for timber to replace lost oars. He did not find what he wanted, so the boats moved on, their progress slowed for the third consecutive day by unusually dangerous sandbars. Darkness came before the exhausted men finally made camp at a place where mosquitoes and ticks were "noumerous & bad." A mile beyond their campsite the next morning, they came ashore and found ash suitable for making oars. While some of the men began felling trees, others made a new towrope using cable Lewis had brought from Pittsburgh. The hunters killed two deer and a bear and also brought back to camp a young horse well fattened by life on the prairies. Clark speculated that the animal had been left behind by an Indian war party. The hunters killed five more deer and another bear as work continued on the oars, but several men were suffering from dysentery, many had boils, and everyone was being attacked by mosquitoes. Lewis's careful preparations had included purchasing mosquito netting in Philadelphia and St. Louis, which he could distribute to the men as some protection, at least at night, from what he called "these most tormenting of all insects."

Mosquitoes would become "emencely noumerous" and "excessively troublesome" later that summer and even more of a plague in the summers of 1805 and 1806. Their bites made the horses nearly frantic and left Lewis's dog howling in pain. To escape their attacks, the men tried greasing themselves by day, standing in the smoke of campfires in the evening, and retreating under netting at night. Swarms of mosquitoes made writing difficult and sleep sometimes impossible. On the return trip, Clark wrote of missing a chance to shoot a bighorn ram because "the Misquetors was So noumerous that I could not keep them off my gun long enough to take Sight." They also inspired one of Clark's best demonstrations of inventive spelling. In the course of the journey, he would find at least nineteen ways to spell the name of "these most tormenting of all insects."[18]

The camp on June 19 was on the south side of the Missouri, near a lake Clark described as being several miles in circumference. It was said to "abound in all kinds of fowls," he wrote, and "great quanties of Deer frequent this Lake dureing Summer Season." He viewed the scene just east of today's town of Lexington with an eye to the future, writing that the land on the north side of the river "is rich and Sufficiently high to afford Settlements," while the lands on the south are "of a good quallity and appear well watered." Despite the tribulations of mosquitoes and ticks, sandbars and sawyers, the expedition's journalists were remaining attentive to the passing scene and faithful in recording their observations. That morning Clark had seen gooseberries and raspberries growing "in abundance" along the banks, and the next day he wrote of seeing pelicans on the river.

Sergeant Ordway, who was usually aboard the keelboat, spent June 21 on shore, hunting with Drouillard. At the end of the day they returned with a deer and a turkey, but Ordway also brought back a description of present-day Lafayette County: "I never Saw as fine Timbered land in my life nor Such Rich handsome bottom land." A week earlier he had observed "Beautiful high Good praries on the South Side," calling the area near present-day Van Meter State Park "the pleasantest place I have ever Seen."

During the crossing of Missouri, the expedition's journalists frequently wrote of "fine timbered land," and they often mentioned specific varieties of trees. In addition to the cottonwoods the crews used to replace broken masts, the ashes that were used for oars, and the willows used for setting poles, at least fifteen other species of trees are named in the Missouri portion of the journals. Clark's entry for June 21, 1804, includes a catalog of the trees in the area and where they grew in relation to the river:

> The Countrey and Lands on each Side of the river is various as usial and may be classed as follows. viz: the low or over flown points or bottom land, of the groth of Cotton & Willow, the 2nd or high bottom of rich furtile Soils of the groth of Cotton, Walnut, Som ash, Hack berry, Mulberry,

Lynn & Sycamore. the third or high Lands risees gradually from the 2nd bottom . . . and are covered with a variety of timber Such as Oake of different Kinds Blue ash, walnut &c. &c. as far as the Praries, which I am informed lie back from the river at some places near & others a great Distance.

Clark was correct in his observations. A note in the Moulton edition of the Lewis and Clark journals adds that "those designations of forest types are accurate and typical of Missouri River topography and forest vegetation."[19]

On June 23, in an area known then as the Fire Prairie, the boats met with winds blowing downriver with such force that visibility was severely limited. As they waited for better conditions, Lewis took time for an inspection of the men's weapons, while Clark left his post in the keelboat and walked along the shore past a wide bend in the river. His day's exploration led to an unexpected night away from the main party: "I Killed a Deer & made a fire expecting the boat would Come up in the evening. the wind continueing to blow prevented their moveing, as the distance by land was too great for me to return by night I concluded to Camp, Peeled Some bark to lay on, and geathered wood to make fires to Keep off the musquitor & Knats."

Two years after his return from the Pacific, Clark, as the Indian agent for the Upper Louisiana Territory, would revisit this area, just east of present-day Independence. At this place he would conclude a treaty with the Osage Indians and supervise the construction of Fort Osage, originally named Fort Clark.

Following his night away from the Corps of Discovery, Clark was picked up early, and the boats moved on, passing a stream known as the Little Blue River. Plums, raspberries, and "wild apples" grew in abundance, while bears had "passed in all Directions thro the bottoms in Serch of Mulberries," which were also plentiful. Deer were seen in large numbers on the prairies and at the riverbanks, where they came to feed on young willows.

After passing the Blue Water River, known now as the Blue River, the boats came to a narrow passage with a dangerous whirlpool on one side. Just beyond that lay a sandbar that twice broke the keelboat's towrope. When the crews finally brought their boats to the mouth of the Kansas River, Captain Lewis decided the expedition would remain in this area for several days. It was June 26. Five weeks had passed since the Corps of Discovery left St. Charles, and the river journey of 390 miles had taken the expedition the breadth of the future state of

William Clark's observation of Carolina parakeets in Missouri was the first documented sighting west of the Mississippi of these now-extinct birds. *Illustration by William T. Cooper, in* Parrots of the World *by Joseph M. Forshaw. By permission of the National Library of Australia.*

Missouri. (The straight-line distance is about 250 miles.) At this point the river turns at a right angle, so when the journey resumed, the boats traveled north for more than two hundred miles before reaching what would become Missouri's border with Iowa.

On June 26, at the future site of Kansas City, Clark wrote that he had "observed a great number of Parrot queets." This is the earliest known sighting west of the Mississippi of the now-extinct Carolina parakeet (*Conuropsis carolinensis*), whose bright green and yellow coloring was once seen from New York to the Rocky Mountains and from the Great Lakes to Florida. About the size of mourning doves, the birds would fly in compact flocks of two or three hundred and stay together even when in danger, a trait that hastened their extinction when they were fired on by farmers whose crops they destroyed. Loss of habitat was another factor, since the birds nested in hollow trees that were cleared as settlement moved west. Carolina parakeets, the only parrots native to the United States, were diminishing in number east of the Mississippi by the 1830s, and the last known specimen died in a zoo in 1918.[20]

The three-day pause at the future site of Kansas City offered the Corps of Discovery relief from the struggles against the Missouri River, but work never stopped. Although the only Indians seen since the expedition left St. Charles were those Clark had arranged to meet and the small party seen by one of the hunters, Lewis ordered the men to build a six-foot-high redoubt of logs and bushes between the two rivers, giving them a protected point of land on which to camp. On checking their boats' cargoes, they found some provisions had been ruined by water. Other items, such as gunpowder and wool clothing that were damp but salvable, were put in the sun. One pirogue was completely unloaded and turned up to dry so that repairs could be made when work on the redoubt was completed. Someone, perhaps Private Silas Goodrich, who was later described by Lewis as "remarkably fond of fishing," caught "Several large Cat fish." The captains used their time to study the area that within seventeen years would become home to Chouteau's Post, a tiny fur-trading establishment destined to be succeeded by the frontier town of Westport Landing and eventually by the modern metropolis of Kansas City.[21]

As they had earlier done at other important confluences, the captains took celestial observations and measured the width of the rivers. Clark recorded that the Missouri was about 500 yards wide and the Kansas about 230 yards at the confluence, with the Kansas being wider above its mouth. Lewis weighed the waters of the two rivers to determine the specific gravity of each, while Clark made a simpler but less pleasant test, concluding that "the waters of the Kansas is verry disigreeably tasted to me."

That same day Sergeant Ordway was quenching his thirst with more refreshing water. "I went out hunting," he wrote, and "passed a fine Spring Running from under the hills I drank hearty of the water & found it the best & coolest I have seen in the country." In his record of the day's events, he added that the Field brothers had brought a young wolf back to camp, "for to Tame," as Ordway explained it. Another of the hunters reported seeing buffalo along the Kansas River, the first sighting of the huge, shaggy-coated American bison by a member of the expedition.

On the last day of their stay at the Kansas River, John Collins, who had repeatedly caused trouble with his drinking, got drunk on whiskey he

"A very large wolf came to the bank and looked at us this morning."
—William Clark, June 30, 1804. *Lynda Stair.*

was supposed to be guarding and invited Hugh Hall to join him in drawing from the barrel. The crews resented anyone stealing from their rations and were, in Clark's words, "verry ready to punish Such Crimes." A court-martial of their peers found both men guilty, with Collins being sentenced to one hundred lashes on his bare back and Hall fifty. The Orderly Book notes that the commanding officers approved the sentences, but Clark wrote in his journal of the party only "inflicting a little punishment," a seeming contradiction.

The boats set out again on the morning of June 29, and when the crews could take their eyes off the river, there were diversions in the last days of the month. One morning a majestic gray wolf *(Canis lupus)* stood at the bank and watched the explorers moving through its territory. When the expedition reached the mouth of the Little Platte River, some of the party ascended this stream and reported finding "Several

Despite the high temperatures, Clark saw deer skipping in every direction.
*Missouri Department of Conservation; Paul Childress, photographer.*

falls well Calculated for mills." In the bottomlands, amidst large quantities of grapes and raspberries, pecan trees were seen. But by the end of June, the men needed neither calendars nor southern trees to remind them of warm weather. In midafternoon on the last day of the month, the thermometers Lewis had brought registered ninety-six degrees, and with a number of his men becoming ill, he ordered a three-hour rest period. In the sultry atmosphere, only the deer remained lively. They appeared in great numbers on the banks, "Skipping in every derection."

Clark's journal entry for the final day of June concludes with the terse statement "Broke our mast." There would be another delay while a replacement was made and installed.

In the first days of July, the expedition passed several reminders of the brief French presence in the region. Across from what is now Leavenworth, Kansas, the men camped on one of a cluster of islands where, according to an engagé, "the french Kept their Cattle & horses . . . at the time they had in this quarter a fort & trading establishment." After dark the next evening, the party made camp across the river from the site of Fort de Cavagnolle, in present-day Leavenworth County, where soldiers had been sent in the mid-1700s to found a post near the Kansa Indians.[22] By 1804 the Kansa village was gone, and only a faint outline on the ground and a few chimney stones remained near the spring where the fort's occupants had obtained their water.[23]

A diorama in the St. Joseph Museum reminds us of the large role the beaver played in the history of the American West. *St. Joseph Museum, St. Joseph, Missouri.*

If the fading vestiges of the old French fort represented an era that had passed, a discovery soon to be made represented an era that was beginning. On July 3, Clark wrote of "a large Pond Containg Beever," his first mention of the North American beaver *(Castor canadensis),* which the expedition would see in far greater numbers as it continued upriver. Two days later, he reported the men "came to for Dinner at a Beever house," adding that Lewis's dog entered the house and drove the beavers out. An encounter about a year later nearly cost the Newfoundland his life. One of the hunters had wounded a beaver, and Seaman was retrieving it from the water when it bit the dog on one leg, severing an artery. For a time Lewis feared he wouldn't be able to stop the bleeding and Seaman would die.

Lewis and Clark's reports of seemingly limitless beaver on the upper Missouri, coupled with the vogue for beaver hats (then recently revived by London dandy Beau Brummell) made the next three decades the golden age of the American fur trade. From St. Louis, the center of that commerce, hardy adventurers set out to test their skills and luck. In the process of creating the legendary figure of the mountain man, they wrote a dramatic chapter in the history of the West, blazed trails countless pioneer families would follow, and brought the North American beaver to near-extinction. The quiet pond the Lewis and Clark Expedition dined beside that day held just beneath its surface the beginning of another American epic.

The Fourth of July, 1804, was welcomed by the Corps of Discovery with the firing of the keelboat's cannon. At sunset the gun was discharged again, and an extra allotment of whiskey was given each man. On the twenty-eighth anniversary of America's independence, the expedition moved upriver fifteen miles through fine landscape, celebrating the occasion by bestowing on previously unmapped streams names appropriate to the day. Entering from the port side was a creek they called "Fourth of July." Another became Independence Creek, the name that a tiny stream at Atchison, Kansas, retains.

Not every name bestowed that day had a patriotic theme. At the Kansas prairie where the men stopped for their midday meal, Joseph Field was bitten on the foot by a snake, leading them to call the place "Jo Fields Snake Prarie." On the Missouri side of the river was an oxbow lake nearly a mile wide and seven to eight miles long, which Clark described as brilliantly clear and containing a large number of geese and goslings. What he named Gosling Lake is believed by some to be the lake that borders Lewis and Clark State Park, located in Buchanan County, between Kansas City and St. Joseph.

North of the lake that Independence Day, the men made camp at a place described by Clark as "one of the most butifull Plains, I ever Saw, open and butifully diversified with hills & vallies all presenting themselves to the river." The description appears with very similar wording in Ordway's account and in Floyd's. When telling Jefferson that seven men were keeping journals, Lewis had added that "in this respect we give every assistance in our power."[24] Clearly the accounts were not written in isolation. Although they are distinct and individual, the journals contain many passages that indicate the men borrowed from each other's writings or talked about what should be mentioned and how to describe it.

Western Missouri, a major migratory flyway, attracted an assortment of waterfowl. Clark wrote of seeing pelicans, a large number of dramatically colored wood ducks with their young, and many geese, whose goslings were "not yet feathered" and unable to fly. After hunting on the Missouri side of the river, George Drouillard told the captains he had seen young swans on a lake. When a nonaquatic bird, a whippoorwill, "perched on the boat for a Short time," a nearby Kansas stream was given the name of Whippoorwill Creek.

On July 7, Clark noticed a "large rat on the Side of the bank," but it is the journal of Patrick Gass that notes "the principal difference between it and the common rat is, its having hair on its tail." Another characteristic, one they were apparently not aware of, is its habit of collecting small, shiny objects such as metal buttons, coins, and nails, a trait that has given it the common names "pack rat" and "trade rat."[25] Several weeks earlier, on May 31, while Lewis was finding "many curious Plants & Srubs," Clark reported "Several rats of Considerable Size was Cought in the woods." In the months ahead Lewis and Clark would provide accounts of more than a hundred species and subspecies of animals not previously described, among them the pronghorn antelope, the prairie dog, the mule deer, the mountain goat, the bighorn sheep, and the grizzly bear. The lowly wood rat, encountered in Missouri, marked the beginning.[26]

In early July, the sweltering heat increased the difficulty of working against the river. Ordway wrote that sweat "pores off the men in Streams," while Clark's journal tells of at least five men being sick from the heat. Since late June, Lewis had ordered three-hour rest periods during the hottest part of the day, but on July 7, Private Robert Frazer suffered an apparent sunstroke. Lewis first bled him and then gave him potassium nitrate, or saltpeter, procedures that the captains reported "revived him much." Although both treatments would have increased the patient's dehydration, Frazer regained his health within a few days.

Near the future site of St. Joseph, the boats passed a beautiful prairie called St. Michael's, described by Clark as having "much the appearance from the river of farms," since it was "Divided by narrow Strips of woods." In the afternoon Sergeant Ordway went ashore to explore this land on the Missouri side of the river. Following a creek until it was too late to get back to the boat that evening, he made camp but found the mosquitoes so bothersome he couldn't sleep. There was, however, some compensation when he related the experience to Clark on his return. "As this Creek is without name," he wrote, "& my Describeing it to my Capt he named it Ordway Creek." The next year the sergeant would have another stream named in his honor, in the dramatic Gates of the Mountains area near Helena, Montana. A later generation would give it the name Little Prickly Pear Creek, while the identity of Ordway's Missouri creek has been lost along with Clark's detailed maps of this area.[27]

On the morning of July 9 the expedition passed a stream flowing into the Missouri from "a large Pond of about three miles in length." Ordway described the pond as "handsom," noting that it contained "a great many beaver, & fish." Located in Holt County, Missouri, it was in a "well timbered" region he called "fine land." Clark wrote that "as our flanking party Saw great numbers of Pike in this Pond, I have laid it down with that name anex'd." Although he indicates he charted the lake, it also fails to appear on his known maps, but the stream described as flowing from the lake was probably the one later called Little Tarkio Creek.[28]

A few miles farther north the expedition passed a prairie on the Kansas side of the river where, according to Sergeant Floyd's account, several "French famileys had Setled and made Corn Some Years ago." He added that during their stay "the Indians came Freckentley to See them and was verry frendley."

The reminder of Indians could have contributed to a period of alarm the expedition experienced that evening. The main party was camped on a point of land on the west bank of the Missouri when the men became aware of another group on the opposite side. Believing this was their hunting party, they signaled a number of times but received no reply. With the fear growing that they could be facing a Sioux war party, the men on the west side of the river fired the keel-boat's gun to warn their hunters of danger and then prepared to repel an attack. After an uneasy night, they crossed the river to investigate and "Soon discovered them to be our men." The hunters, it turned out, had fallen into a deep sleep early in the evening and were unaware of the concern they had caused.

They had reason to be exhausted. Despite the muggy heat and tormenting insects, they were expected to bring in enough game to feed about forty-seven hardworking men. Part of Clark's entry for July 9 reads: "Sent one man back . . . to mark a tree, to let the party on Shore See that the Boat had passed." His notation is a reminder of another burden placed on the hunters. They not only had to shoot game and then, with the help of their horses, get it back to the river, but they also had to gauge quite accurately how many miles the boats would travel during their absence so that a rendezvous could be made.

Realizing that both his hunters and boatmen were fatigued and suffering from the heat, Lewis decided July 12 would be a day of rest. The permission to rest came too late for Alexander Willard. He was charged with lying down at his post and sleeping while on guard duty the previous night, a capital offense since it endangered the lives of the other men. Because the death sentence could be imposed in this case, the captains constituted the court. After hearing Willard admit to lying down but not to sleeping, they found him "guilty of every part of the Charge" and sentenced him to one hundred lashes to be imposed in equal portions on four successive evenings.

Near the end of his long entry for that day, Clark wrote: "Tred [tried] a man for sleeping on his Post & inspected the arms amunition &c. of the party found all complete." At the time or probably later, he interlined the letters "Wld," an abbreviation for *Willard*. Despite the serious nature of the offense under military law, Clark's journal seems to downplay Willard's falling asleep. Clark does not spell out the man's name, and he includes the court-martial as one of several events related in a single sentence. And, as with the previous cases of Hall and Collins, there is nothing in the journals to indicate whether Willard's punishment was fully carried out.

Ordway wrote that some of the men used this day of rest to wash their clothes. Drouillard brought his two-day total of deer to eight. Lewis took more celestial observations, and there was the usual scouting to determine the quality of the land. Soon after breakfast, Clark left with five men in one of the pirogues to ascend the Nemaha River, which enters the Missouri near the point where the southeastern tip of Nebraska meets the state of Missouri. About three miles up the Nemaha, in present-day Richardson County, Nebraska, he went ashore to study several Indian burial mounds. After admiring the surrounding prairie, he returned to the pirogue for the trip back to camp, stopping en route to gather some grapes and, he said, to inscribe his name, along with the day, month, and year, on a sandstone bluff where he observed Indians had earlier left inscriptions.

On July 14, just below the Nishnabotna River, which flows through Atchison County, Sergeant Ordway reported: "We Saw three large Elk the first wild ones I ever Saw. Capt. Clark & drewyer Shot at them, but the distance was too long." It was not Clark's only frustration that day. The keelboat and pirogues had been caught by a sudden and violent squall hours earlier, and Ordway's report added: "Capt. Clarks notes & Remarks of 2 days blew Overboard this morning in the Storm, and he was much put to it to Recolect the courses."

The next morning, July 15, Ordway accompanied William Clark on a shore excursion in the area that is today Nemaha County, Nebraska. The elk they hoped to find eluded them, but they did see large quan-

The Lewis and Clark Expedition crossed Missouri in mid-May to mid-July, when prairie flowers such as phlox would have been blooming. *Ann Rogers.*

tities of grapes, cherries, plums, and berries of various kinds. From atop a high ridge the two men had a view of open plains to the west, Ordway wrote, "as far as our eyes could behold." Looking back across the river to the region that would be the northwestern tip of Missouri, they saw an "extensive prarie," which they found to be "verry handsome."

The expedition's last full day in Missouri on the outward journey was spent on Bald-pated Prairie, located in what is now Atchison County. Throughout most of the day Lewis rode along the Nishnabotna River, which flowed parallel to the Missouri through land he described as "one of the most beautiful, level and fertile praries that I ever beheld." To the east were the Bald Hills, a long ridge that formed the backdrop for a variegated landscape of open expanses and tree-lined banks that Lewis found "handsome country." Bald-pated Prairie held rewards for the hunters and fishermen as well. Clark's field notes record that "G Drewyer kill'ed 3 deer, & R Fields one," while Goodrich "caught two verry fat Cat fish."

On July 18, 1804, the Corps of Discovery set out in fair weather, and in the course of that day's travel the boats moved beyond what is now the northern boundary of Missouri. The traverse of the area that would become Missouri had taken sixty-six days and had advanced them along nearly six hundred miles of the Missouri River. Far ahead lay the Rocky Mountains, the Pacific shores, and, finally, a triumphant return to St. Louis.

# 3

## THE MONTHS BETWEEN

---

T hree days after leaving the region that would become Missouri, the expedition reached the mouth of the Platte River, flowing in from the west. For St. Louis–based traders, who had not yet traveled beyond the Mandan villages, the Platte was seen as the dividing line between the lower Missouri and the upper portion of that river. For Lewis and Clark, arriving at the Platte presented an opportunity to hold a council with the first Indians they had seen since Clark's pre-arranged meeting just a few miles beyond St. Charles. Lewis decided they would remain for several days to see if any chiefs could be found. While waiting, he and Clark used the time to update their maps and notes. The delay failed to produce a meeting with Indians, and when none were found, the party moved about ten miles north to try again. Scouts sent by Lewis to a nearby village returned saying the place was deserted, apparently because at this season the Indians were away hunting buffalo. Sergeant Floyd's journal entry for July 25 begins: "Continued Hear as the Capts is not Don there Riting." He then added: "Ouer men Returnd whome we had Sent to the town and found non of them at Home."

Several days later Drouillard brought to camp a Missouri Indian who was living with the Oto tribe and was willing to act as a liaison. On August 2, about fifty miles above the Platte, Lewis and Clark welcomed to the place they called Council Bluff a group of Otos and Missouris, among them a number of lower-echelon chiefs.[1] The format for the council established what would become a ritual as the Corps of Discovery met successive tribes in the months ahead. There were speeches by the captains stressing the sovereignty of the United States, promises of trade, calls for intertribal peace, and speeches by the Indians. Drawing on goods purchased by Lewis and systematically arranged and packed at Wood River, the captains would bestow gifts, both practical and symbolic. Among the symbolic gifts Lewis brought from Philadelphia were peace medals, with the image of President Jefferson or Washington on one side and an image such as hands clasped in friendship on the other. Chiefs took pride in these medals and in the certificates Lewis would present after filling in the individuals' names.

While the expedition waited for Indians at the place called Council Bluff, Sergeant Floyd wrote: "I am verry Sick and Has ben for Somtime." For a while he seemed to recover, and his brief journal entries for the next eighteen days make no mention of the illness. Then suddenly his condition became critical. On August 20, as the keelboat neared the present site of Sioux City, Iowa, it was clear Floyd was dying. Ordway's journal gives an account of what followed:

> Sergt. Charles Floyd Expired directly after we halted a little past the middle of the day. he was laid out in the Best Manner possable. we proceeded on to the first hills. . . . there we dug the Grave on a handsome Sightly Round knob close to the Bank. we buried him with the honours of war. the usal Serrymony performed by Capt. Lewis as custommary in a Settlement, we put a red ceeder post . . . & branded his name date &C— we named those Bluffs Sergeant Charles Floyds Bluff.[2]

When the expedition had moved far beyond his solitary grave, his cousin Nancy received a letter from her brother telling her: "Our dear Charles died on the voyage. . . . He was well cared for as Clark was there."[3] Throughout the final hours of the sergeant's life, Clark and others had sat with him, offering comfort but unable to do much more. If Floyd's death was due to a perforated or ruptured appendix, as is generally believed, the finest doctors of the time could not have saved him, because no successful appendectomy was performed until the 1880s.[4] The twenty-two-year-old Kentuckian recruited by Clark and praised by Lewis as "a young man of much merit" had become the first U.S. soldier to die west of the Mississippi.

A week after the death of Floyd, with Patrick Gass chosen to replace him as sergeant, the expedition located a group of Yankton Sioux. The meeting went so well that the captains asked Old Dorion to remain with them to encourage intertribal peace and make arrangements for some of their chiefs to visit President Jefferson. A meeting in September with the Teton Sioux proved far more challenging. The tribe had previously harassed both rival tribes and European traders, and these Indians quickly applied their bullying tactics to the Corps of Discovery. A chief who had come aboard one of the boats jostled and insulted Clark while insisting the expedition's flotilla would not be allowed to proceed. Clark drew his sword, Indians on the banks prepared their bows and arrows, and Lewis ordered his men to ready their guns. After an

A hundred-foot-tall obelisk at Sioux City, Iowa, marks the grave of Sergeant Charles Floyd, the only member of the Corps of Discovery to lose his life on the journey. *Ann Rogers.*

exchange of words, the standoff ended. But throughout four uneasy days, periods of communication and hospitality alternated with misunderstandings and threats. When the Arikaras were encountered in early October, they proved as gracious as the Tetons had been belligerent, and the visit with them was entirely amicable.

While the Corps of Discovery was moving toward Sioux territory, there had been growing concern that nineteen-year-old George Shannon, the expedition's youngest member, was missing. The relatively inexperienced hunter had become separated from Drouillard and then pushed ahead of the boats for more than two weeks, mistakenly thinking he had fallen behind them. He later told Ordway of seeing what were apparently Indian tracks, which he had supposed were made by men of the expedition. When they finally caught up with him, they learned he had spent his bullets and then gone hungry for twelve days except for a few grapes and one rabbit he had shot by cutting some sticks and putting them in his gun. One of his horses had given out and been left behind, and he was preparing to kill the other for food. Clark wrote in dismay that the "man had like to have Starved to death in a land of Plenty for the want of Bulletes or Something to kill his meat."

As Clark had traced the progress of spring at Camp Dubois, he now recorded the coming of winter to the Northern Plains. On September 19 he wrote that "the leaves of some of the cottonwood begin to fade." October 5 brought the first "slight white frost," and on October 9 he watched "geese passing to the south." By October 14 "the leaves of all the trees . . . except the cottonwood" had fallen, and three days later he wrote that "antilopes are passing to the black hills to winter, as is their custom." The first snow fell before the men reached, on October 26, the place where they would spend the winter of 1804–1805.

The site the captains selected was near the Mandan villages, about fifty miles north of present-day Bismarck, North Dakota. The villages, with a combined population of over four thousand, included Indians from the Mandan and Hidatsa tribes, who had come together as a defense against the Teton Sioux. Near the southernmost of the five villages, the Corps of Discovery erected a log structure later called Fort Mandan. Although the fort was not completed until Christmas Eve, harsh weather forced the men into the partially built shelter in mid-November.

During the bitter winter at Fort Mandan, with the temperatures sometimes reaching forty degrees below zero, the captains were called upon to treat frostbite among their own men and the Indians. The long months of severe cold were difficult for the hunters and tedious for the men assigned routine chores in the confinement of the fort. For Lewis and Clark, however, one of the most important activities of the five-month encampment was learning as much as possible about the rivers and lands to the west, because the maps they had brought from St. Louis showed little between the Mandan villages and the Pacific coast. From the Hidatsas, whose forays carried them as far as the Continental Divide, the captains were able to gain information about the Great Falls of the Missouri, the division of that river into three forks, and possible land routes through the Rocky Mountains.

At the end of February, Lewis had the two pirogues chopped free of the ice and sent a detail of men to cut cottonwoods to make dugout canoes. These would replace the keelboat, which was considered too large for the upper Missouri and was being readied for its return to St.

Reconstructed Mandan earth lodges stand today at Fort Lincoln State Park, near Bismarck, North Dakota. *Ann Rogers.*

Louis. Aboard it would be more than a hundred plant and mineral specimens Lewis had collected and labeled for identification. They would be sent to President Jefferson, together with buffalo robes, "highly embellished with Indian finery," various animal skins, and the horns of elk, mountain rams, and mule deer. An ear of Mandan corn was packed, along with examples of Mandan pottery. Cages would hold the live animals: a prairie dog, a grouse, and four magpies. Clark was preparing the papers he would send, including a detailed map of the lower Missouri and a chart of the region's Indian tribes. As the captains assembled documents, letters, and one of the journals, Clark wrote: "We are all day ingaged packing up Sundery articles to be Sent to the President of the U.S."

Under the command of Corporal Richard Warfington, whom the captains had found to be trustworthy, four other members of the Corps of Discovery and several French boatmen would return the keelboat, joined by a number of chiefs who had accepted an invitation to visit the president. After leaving Fort Mandan on April 7, the keelboat arrived at St. Louis on May 20, 1805, a year to the day, less one, from the May afternoon when the expedition set out from St. Charles. Pierre Chouteau greeted the chiefs in St. Louis and then escorted them to Washington, while Amos Stoddard took charge of forwarding to Jefferson the keel-

Eight weeks after giving birth to a son, Sacagawea set out with the expedition, carrying the child on her back. *Ann Rogers.*

boat's cargo. (Although all the animals survived the trip to St. Louis, only the prairie dog and one magpie eventually reached the nation's capital. Jefferson later received a letter at Monticello telling him the bird and "little burrowing dog" both "appear healthy" and had been put in the room where he received callers.)[5]

During the winter at Fort Mandan, the captains hired the services of Toussaint Charbonneau, an interpreter who had lived for years in the Northwest and claimed the ability to communicate with tribes they would meet along the upper Missouri. The captains were interested to learn that his young wife, who was expecting her first child, had lived near the headwaters of the Missouri until raiding Hidatsas kidnapped her from her Snake, or Shoshone, tribe when she was about ten.[6] The Shoshones, as the captains knew, owned large numbers of horses, and the expedition would need horses for the overland crossing of the Rockies. Sacagawea could be helpful in locating her tribe and in the negotiations with them. Moreover, she and her child would serve as a sign to the Indians that the explorers came in peace, because to the Indians a war party accompanied by a woman was unthinkable.

On the eleventh of February, 1805, the men waited out her long and difficult labor until, in the words of Sergeant Gass's journal, she "made an addition to our number." The arrival may have been speeded by another interpreter's suggestion that she be given crushed rattles from a snake. Lewis wrote: "Having the rattle of a snake by me I gave it to him and he administered two rings of it to the woman broken in small pieces with the fingers and added to a small quantity of water." On learning she delivered the child no more than ten minutes later, Lewis observed: "This remedy may be worthy of future experiments." Eight weeks later, with her baby son strapped to her back, Sacagawea joined her husband and the Corps of Discovery as the voyage resumed.

The same afternoon that the keelboat started downstream toward St. Louis, the thirty-three members of the permanent party, traveling now in two pirogues and six smaller boats, set out on the Missouri to continue westward. Following the pattern established early in the journey, Clark would usually remain with the boats while Lewis would often choose to walk along the banks, accompanied by his dog, Seaman. On the Northern Plains the party saw "immence quantities of game in every direction . . . consisting of herds of Buffaloe, Elk, and Antelopes with some deer and woolves." But Lewis noted: "We only kill as much as is necessary for food."

A week after leaving Fort Mandan, the explorers began to see the "white bear," the humpbacked grizzly, whose awesome size and strength they would come to fear even more than hostile Indians. Meriwether Lewis provided detailed descriptions of this animal that was far larger and more dangerous than the black bears the expedition found in Missouri. As the party moved through the Northwest, Seaman made his contribution by barking to signal the approach of these dreaded beasts. When the bears neared the expedition's campsites, Lewis wrote, "our dog gives us timely notice of their visits, he keeps constantly padroling all night."

On a day when the men were still shaken by an encounter between six hunters and a grizzly who remained in pursuit after taking eight bullets, there was a second brush with disaster. Having judged the white pirogue the most stable of the boats, Lewis had stored in it the captains' "papers, Instruments, books, medicine, a great part of our merchandize and in short almost every article indispensibly necessary to . . . insure the success of the enterprize in which we are now launched to the distance of 2200 miles." It also held several persons who could not swim. Charbonneau, described by Lewis as "perhaps the most timid waterman in the world," was relieving Drouillard at the helm when a sudden gust of wind caught the pirogue.

Charbonneau first turned the rudder the wrong way, then lost hold of it and panicked as the boat turned on its side and began to fill with water. While Lewis watched in horror from the shore and shouted orders that couldn't be heard, experienced riverman Pierre Cruzatte told crewmen to bail the water and threatened to shoot Charbonneau if he didn't get control of the rudder. As two men used kettles to bail, Cruzatte and two others rowed the sinking vessel to shore. Meanwhile, Sacagawea, sitting in the rear of the boat, maintained her composure and gathered from the water as many floating articles as she could retrieve. Within a week the captains named a "handsome river" in her honor. Although a later generation, unfamiliar with Lewis and Clark's maps and journals, called it Crooked Creek, the name Sacagawea River has since been restored to the eastern Montana stream.[7]

Late in May the boats reached the Missouri Breaks, an area of white sandstone cliffs sculptured by water and wind into fantastic formations Lewis called "seens of visionary inchantment." Even today, this portion of central Montana is one of primitive beauty, nearly untouched by man.

Just beyond the breaks, Lewis would name a river for his cousin, Maria Wood, but before the name was given, a crucial decision had to be made: Was that river or the one running to the south the Missouri? Nearly all of the party decided quickly that the northern branch, the Marias, was the Missouri. Lewis and Clark thought otherwise, but they were puzzled by the volume of water in the Marias and by the Hidatsas' failure to tell them about so large a river. After the captains spent a week examining the choices, they finally passed by the Marias and continued on the river they believed was the Missouri. On the thirteenth of June, their choice was confirmed when Lewis and his advance party of four men heard the tremendous roar of the Great Falls.

The thunderous cascades that Lewis heard from a distance of seven miles sent plumes of spray into the air above a series of five falls, the first and largest being three hundred yards across and more than eighty feet high. Although he was awed by the "sublimely grand specticle," his thoughts soon turned to the difficult portage these falls made necessary. Clark, on his arrival, surveyed an eighteen-mile route and set up stakes as guides, marking as short a traverse as possible over the uneven terrain carpeted with prickly pear cactus. The red pirogue and some supplies had been hidden at the Marias; the white pirogue

Lewis described the Great Falls of the Missouri, near present-day Great Falls, Montana, as "truly magnifficent and sublimely grand." The scene is still impressive, despite a dam above the falls and far less water cascading over the rock ledge. *Ann Rogers.*

and more supplies were cached near the falls. Two crude wagons built to haul the canoes and heavier gear would have to be pulled by the men. Other supplies would be carried on men's backs. The twelve-day portage was accomplished despite intense heat, miles of walking on moccasined feet pierced by cactus, a flash flood preceded by hail-stones large enough to knock a man to the ground, and the constant threat of grizzlies and rattlesnakes.

While the portage was under way, Lewis began supervising the assembly of a boat whose thirty-six-foot iron frame he had brought from Harpers Ferry for use on the upper Missouri. His men cleaned it of rust, bolted the frame together, cut timber for the struts, and sewed animal skins to cover the hull. Lewis proudly noted that his experimental boat looked "extreemly well" and "will be very light, more so than any vessl of her size that I ever saw." His pride turned to dismay when the seams began to leak soon after it was put in the water. No pitch or tar was available for caulking, and Lewis's mixture of buffalo tallow, beeswax, and charcoal had proved inadequate. With more than two weeks invested in the project, Lewis abandoned his "favorite boat" and had a pair of canoes built to replace it, a task requiring an additional five days at the portage site.

On July 4, 1805, one year after the Corps of Discovery celebrated Independence Day in northwest Missouri, Lewis wrote in his journal that he and Clark had decided not to send a detachment back to St. Louis from the Great Falls, a plan they had earlier considered but "never once hinted to any one of the party." With the journey proving longer, more arduous, and even more uncertain than anticipated, they wanted every man for whatever lay ahead.

The stay at the falls had extended to a full month. During the portage a number of men became ill, others suffered injuries, and almost all were exhausted, leading Clark to write: "To State the fatigues of this party would take up more of the journal than other notes which I find Scercely time to Set down." One hopeful note was Sacagawea's recovery from an illness that nearly took her life at the start of the portage. Along with concern for the woman and the "young child in her arms," there was the realization she was "our only dependence for a friendly negociation with the Snake Indians on whom we depend for horses to assist us in our portage from the Missouri to the columbia River."

Increasingly concerned about finding Shoshones, they set out again on July 15. Clark and three other men were making their way over an old Indian road when the rest of the party, traveling now in eight canoes, entered a canyon Lewis called the "gates of the rocky mountains." North of Helena, Montana, the area is known today as the Gates of the Mountains. "Clifts rise from the waters edge on either side perpendicularly to the hight of 1200 feet," he wrote, while "the river appears to have forced it's way through this immence body of solid rock for the distance of 5¾ miles." Traveling the narrow corridor in late evening light, Lewis felt "every object here wears a dark and gloomy aspect" as the "projecting rocks in many places seem ready to tumble on us."

By July 25, Clark and his weary party, unsuccessful in their search for Shoshones, reached the headwaters of the Missouri River, where they were joined within two days by Lewis and the boat crews. Three streams converged here, but only one would further their advance to the Pacific. After studying the possibilities independently, both Lewis and Clark decided it was the river they named the Jefferson that would carry them westward.[8] Sacagawea, who had been kidnapped years earlier in the Three Forks area, was unable to tell the captains

Today's visitors can take an afternoon boat trip through the Gates of the Mountains near Helena, Montana, seeing the area Lewis described in 1805. *Ann Rogers.*

which river to follow, but ten days later she recognized a large rock formation having a shape somewhat like a swimming beaver and said the "beaver's head" was a landmark to her people, who would be nearby at this season.

Encouraged, Lewis took Drouillard, John Shields, and Hugh McNeal and set out on foot to find Shoshones. On August 12 they followed an Indian trail across 8,000-foot Lemhi Pass on the present Montana-Idaho border, becoming the first U.S. citizens to cross the Continental Divide. Lewis was overjoyed when he "first tasted the water of the great Columbia river." (The stream he drank from flowed to the Lemhi River, then to the Salmon and Snake Rivers, and finally to the Columbia.)[9] His joy increased the next day when he found Indians who directed him to their village. With Drouillard to help bridge the language barrier, Lewis met with Chief Cameahwait and learned that despite their horses the Indians were without meat and subsisting on berries. Realizing that any game his men could provide would facilitate the negotiations, he was glad Drouillard was also his best hunter.

At Lemhi Pass, on the Montana-Idaho border, Lewis and three of his men became the first U.S. citizens to cross the Continental Divide. *Ann Rogers.*

Although the Indians provided Drouillard and Shields with horses, these proved no match for the speed and agility of a herd of antelope, the principal game animal in the region. McNeal mixed a paste of flour with some berries to form a "new fashoned pudding" the chief declared "the best thing he had taisted for a long time." Lewis, accustomed to heartier fare, remained "hungary as a wolf."

Meanwhile, Clark and the rest of the party were traveling through country Sacagawea had last seen as a child. Upon their arrival at the village, she was brought to the council to act as interpreter. In this meeting, she gave the expedition one of its most dramatic and improbable moments when she recognized Chief Cameahwait as her brother. "She jumped up, ran & embraced him, & threw her blanket over him & cried profusely."[10] Other emotional scenes followed. She learned most of her family was dead, but she met surviving members and was reunited with her childhood friend who had escaped when Sacagawea was captured. Aided no doubt by this remarkable meeting of Sacagawea and her brother, the captains were able to obtain from the Shoshones about thirty horses.

In discussions with the Indians, Lewis and Clark were told the river into which the Lemhi emptied could not be traveled, owing to its fast current, dangerous rapids, and the scarcity of game along its banks. But to the captains a river flowing westward through the mountains offered a possibility they could not easily pass by. While Lewis cached more supplies, supervised the repacking of baggage, and continued bartering for more horses, Clark set out with a contingent of men along the river he named for Meriwether Lewis. His reconnaissance soon confirmed the Indians' reports. What he found was mile after mile of white water coursing between steep and rocky banks. The river is now called the Salmon, and one treacherous portion Clark saw carries a name he would have understood: the River of No Return. Four days after starting his exploratory probe he decided that travel by this route would be impossible.

Having accepted the land route as their only alternative, the captains hired an Indian guide known as Old Toby to lead them over the mountain trail used by the Nez Percé Indians in their annual crossing of the Bitterroots. On August 30 the expedition began moving northward. Four days later there was a portent of what lay ahead when a cold rain turned to sleet, glazing the rocky precipices on which they were traveling. Better fortune awaited them on a meadow where they met friendly Flathead Indians and obtained from them more than enough horses to replace those injured in falls during the previous day's storm. The expedition then continued northward through a valley presided over by the magnificent and formidable Bitterroot Range. At Travelers Rest, near present-day Missoula, Montana, the men made camp and prepared to cross these mountains.

The eleven-day traverse of the Bitterroots, over an Indian road later known as the Lolo Trail, was the expedition's worst ordeal. The trail along the ridges of the mountains was not clearly defined, and in this heavily timbered region fallen trees often blocked the path. Old Toby, it turned out, had little familiarity with the route, and when snow obscured the trail, he led the party far off course. The horses had trouble keeping their footing; some were injured in falls, while others simply gave out. Game, scarce before, virtually disappeared. To provide food, Lewis decided they would kill the horses that had the least value as pack animals. Clark's journal entry for September 15 ends: "We melted the Snow to drink, and Cook our horse flesh to eat."

September 16 brought another early winter storm, and by evening eight inches of new snow obliterated the trail, which had now reached

Today, a two-lane road parallels the Lolo Trail, where the explorers struggled for eleven days to make their way through the mountains. *Ann Rogers.*

an elevation of 7,000 feet. Travel, slow before, became arduous. Everyone was cold, wet, weak from hunger, and exhausted. The possibility grew that the expedition would fail and that they would all perish in the Bitterroots. In hopes of "reviving ther Sperits," Clark took six men and pushed out ahead in search of level country with game they could kill and send back to those still working their way out of the mountains. On September 20, Clark and his group passed a final ridge and emerged into a beautiful prairie in present-day Idaho, where they were joined two days later by the rest of the expedition. They had crossed the Rockies.

With the Bitterroots behind them, they rested and tried to regain their strength. Nez Percés brought them food, but the unfamiliar diet of camas roots and dried fish made the captains and many of their men violently ill. Some who were well enough began work on the five canoes needed for the next stage of the journey, work made easier by following the Indian method of burning out rather than carving out each log's interior. After about two weeks with the Nez Percés, the captains branded their remaining horses, which they planned to use for the return trip, and left them in the care of this hospitable tribe. In 1892, a branding iron with the marking "U S Capt. M. Lewis" would be found on a sandbar in the Columbia River, and it is now in the possession of the Oregon Historical Society.[11]

Accompanied by two Nez Percé guides, the expedition set out again by water, making its way down the Clearwater and the Snake to the Columbia. The men of the Corps of Discovery had, weeks earlier, become the first U.S. citizens to cross the Rocky Mountains, and on October 16, 1805, they became the first white men to see the Columbia River east of the Cascade Range.[12] The Pacific once more seemed an attainable goal, but reaching it would not be easy. When rapids forced portages, the men who were unable to swim carried the expedition's guns and ammunition, the captains' papers, and other valuables overland while other men guided the canoes past the most hazardous portions of the river.

"Ocian in view! O! the joy!" Clark's journal entry for November 7 proclaims: "We are in View of the Ocian . . . this great Pacific Octean which we [have] been So long anxious to See." But what the explorers were actually seeing was the wide Columbia estuary; the long-awaited Pacific was still a frustrating week away. Rain fell almost constantly, soaking everyone, while choppy waters caused some to become seasick. At times, strong head winds made forward movement impossible, and at night, Ordway complained, the dispirited travelers "had Scarsely room for to camp" between the steep hills and the encroaching tides.

When they at last beheld "with estonishment the high waves dashing against the rocks & this emence ocian," it was mid-November and time to construct winter quarters. On the evening of November 24, Lewis and Clark asked each adult member of the party, including York, a black slave, and Sacagawea, an Indian woman, to express a choice regarding the fort's location. Essentially, the choice was between finding a site a short distance inland on the river or one south of the Columbia and near the coast. Clark listed thirty names and recorded and tallied the responses, including Sacagawea's preference for "a place where there is plenty of Potas."[13] The vote favored examining the south side, which was also the captains' choice. About four miles southwest of present-day Astoria, the explorers built Fort Clatsop, a replica of which stands today on the site selected in 1805.

Christmas Day brought an exchange of small gifts and an attempt at festivity, though the dinner consisted of "pore Elk, So much Spoiled

Part of the reconstruction of Fort Clatsop, near Astoria, Oregon. The explorers moved into their winter quarters on Christmas Eve, 1805. *Ann Rogers.*

that we eate it thro' mear necessity." As weeks passed, the hunters had to range farther to find game and had greater difficulty bringing it back. The winter at Fort Clatsop grew increasingly dreary, with rain and fog so prevalent that the journals record only six clear days during the four-month stay. The constant dampness permeated firewood, spoiled meat, rotted clothing, damaged equipment, caused illness, and created a sense of confinement that tested tempers and lowered spirits.

Throughout the winter a rotating detail of several men was assigned the tedious chore of boiling countless gallons of sea water to obtain the salt needed to preserve meat on the eastward trip. (A reconstruction of the salt cairn can be seen today at Seaside, Oregon, about twelve miles south of Fort Clatsop.) At the fort, the captains used the time to expand and refine their notes and maps, just as they had used the winter at Fort Mandan to work on data gathered in the first months of the journey. Men stood guard duty to protect diminished supplies from Indians who visited almost daily, while other men made clothing, including more than three hundred pairs of moccasins. These would be needed for the return journey, a journey everyone was impatient to begin.

Cannon Beach, on the Oregon Coast. Rain and fog were so common during the winter encampment that the expedition's journalists recorded only six clear days. *Ann Rogers.*

Finally, in late March, Lewis and Clark turned over the fort to a Clatsop chief, from whose tribe the structure derived its name, and the Corps of Discovery prepared to leave this place where they had lived, in the words of the captains, "as comfortably as we had any reason to expect." They did not leave their problems behind. Traveling east on the Columbia meant traveling against that river's strong current, a voyage made more difficult by Indians who pilfered continually and even stole Seaman. The thieves quickly surrendered the dog when they realized the Corps of Discovery considered this a serious offense.

The explorers had been so eager to leave Fort Clatsop that they arrived back at the Lolo Trail much too early in the spring for a crossing of the Rockies. Almost six weeks passed before they could begin their eastward traverse of the Bitterroots, a long wait for those "as anxious as we are," Lewis wrote, "to return to the fat plains of the Missouri and thence to our native homes." The snows were still deep in the mountain passes when the expedition set out, but this traverse was accomplished in only six days compared to the eleven required for the westward crossing.

At Travelers Rest, reached on June 30, the captains divided the party. Both groups would later divide again, but all were to assemble at the confluence of the Yellowstone and the Missouri. Clark's group, which included the Charbonneau family, would travel through Shoshone country to the Three Forks and then along the Yellowstone River.

Clark's group experienced no major difficulties, but Lewis's party was less fortunate. On July 26, after a long reconnaissance of the Marias, Lewis and the three men accompanying him encountered eight Blackfeet, with whom they uneasily camped for the night. Towards dawn the Indians attempted to steal their guns and horses, and in the ensuing struggle one Indian was stabbed and another was shot. Taking four of the best horses, Lewis and his companions rode all day and throughout the night, almost without pause, more than one hundred miles to the Missouri River, where they were relieved to find the rest of their party.

One more misfortune occurred before the reunited group joined Clark. While pursuing an elk in a thicket of trees, Lewis was accidentally shot in the thigh by Pierre Cruzatte, a better waterman than hunter, who was limited by blindness in one eye. Although the ball missed both bone and artery, the painful wound left Lewis feverish that night and restricted his mobility for several weeks. When his party caught up with Clark's, Lewis cited discomfort from his injury and the difficulty of trying to write while lying on his stomach as the reasons to "leave to my frind Capt. C. the continuation of our journal."

By mid-August, after a separation of almost six weeks, all members of the expedition were reunited and moving rapidly along the Missouri River towards the Mandan villages. On their arrival there, Private John Colter asked the captains to release him so he could join two trappers headed west to the Yellowstone. Colter had proved an able and valuable member of the Corps of Discovery, and since his services were no longer crucial, the two leaders were quick to grant his request and wish him well in his future endeavors.

Farewells were also said to Sacagawea, Charbonneau, and the couple's son, whom Clark had nicknamed "Pomp." Clark offered to take the three to St. Louis, but for the time being they would remain at the Indian villages. Joining the returning party was a Mandan chief, Sheheke (also called "Big White"), who had accepted Lewis's invitation to visit President Jefferson. With his wife and son, Sheheke would travel to St. Louis in preparation for his trip to Washington.

Arikara chiefs declined a similar invitation when Lewis and Clark held a council with them four days after leaving the Mandan villages,

and a week later there was an unpleasant exchange with Teton Sioux;
but the Corps of Discovery's final encounter with Indians was with a
small group of Yankton Sioux, whose friendly reception repeated the
experience the explorers had with this tribe during the outward voyage.

Before all thoughts could turn to home, there was one remaining
mission. At noon on September 4, the two captains and a number of
their men climbed a bluff rising from the east side of the Missouri
River and located the grave of the one member of the Corps of Discovery who lost his life on the journey. Finding the burial site dis-
turbed, they restored it and then said a last goodbye to Sergeant
Charles Floyd, who had not been with them when they saw the jagged
peaks of the Bitterroots or the white-crested breakers on the Pacific
shores but who had found much to admire in the gentler landscapes
of Missouri.

# 4

# THE RETURN THROUGH MISSOURI

O n September 9, 1806, the Corps of Discovery passed the mouth of the Platte River and continued rapidly downstream into the area that would become Missouri. After a day's journey of seventy-three miles, the men halted for the night opposite Bald-pated Prairie, which Lewis had described on their upriver voyage as "one of the most beautiful, level and fertile praries" he had ever seen. As Clark noted on their return, the prairie was the place they had camped July 16 and 17 in 1804. From the charts and maps he made on the westward crossing, he knew dropping below the Platte River put the expedition within six hundred miles of its starting point, a distance that could be covered in days. And on that September evening opposite Bald-pated Prairie, Clark could measure progress in another way. A month after the accidental shooting he was finally able to write: "My worthy friend Cap Lewis has entirely recovered . . . and he Can walk and even run nearly as well as ever he Could."

In the next few days the men became aware of changes in terrain, wildlife, and climate. Tall timber was more evident, especially oak, elm, hickory, and walnut, and in these wooded areas the hunters found a great number of deer, as well as raccoon and wild turkey. The nights suddenly became warm compared to those in the Northwest, and blankets that had recently been a necessity were cast aside. A welcome change was the respite from mosquitoes, which had been a torment since the Great Falls but were "no longer troublesome on the river" south of the Platte.

Another contrast was the number of boats they were meeting. In the one thousand miles between the Mandan villages and the Platte, the expedition saw only three trading parties, but during the six-hundred-mile crossing of Missouri at least ten groups were encountered. The day after returning to Missouri, the captains spent a half-hour talking with four Frenchmen from St. Louis, who were in a pirogue headed for the Platte River to trade with the Indians. When the traders offered anything they had on board, the captains, according to Clark, accepted "a bottle of whisky only which we gave to our party." Just three miles farther on, they met a larger pirogue, this one also from St. Louis and carrying seven men on a trading expedition to the Omahas. The conversation with the second group was brief and overshadowed by the difficulties the river presented.

While they had been able to cover sixty-five miles that day, it had been along "a very bad part of the river" with "moveing Sands and a much greater quantity of Sawyers or Snags than above." Steering clear of these hazards was made more difficult by the low state of the water. Sandbars and sawyers had plagued the expedition in the first crossing of Missouri, but low water had not been a problem. When the keelboat and pirogues were moving westward through Missouri in 1804, the river was full with snowmelt and spring rains. Now, in late summer, the flow of the Missouri and its tributaries was significantly diminished. The Nemaha River "did not appear as wide as when we passed up," Clark noted, while the "Wolf river Scercely runs at all."

The creatures for whom this second stream was named made their presence known after dusk. "Wolves were howling in different directions this evening after we had encamped," Clark wrote, as the men paused on an island north of present-day St. Joseph. Joining their voices to the nocturnal chorus were coyotes (*Canis latrans*), or "little prarie wolves" as he called them. The Corps of Discovery had listened to them throughout much of the journey, and the captains had been the first to provide a detailed description of them. Clark admitted that when he heard these animals barking west of the Rockies he had mistaken them for "our Common Small Dogs." Even now, most of the men "believed them to be the dogs of Some boat . . . which was yet below us."

Boats were, in fact, coming upriver toward them, and the next day they saw a familiar face. They had gone about seven miles when they came upon a pair of pirogues from St. Louis. One, loaded with trade goods from the Chouteau warehouse, was headed for the Platte; the other carried trappers who planned to go farther upriver. Among them was one of the engagés who had accompanied the Corps of Discovery to the Mandan villages in 1804. Although Clark's journal and Ordway's mention the chance encounter, neither gives any hint of the Frenchman's reaction to seeing again the explorers with whom he had spent almost a year. If he expressed the slightest curiosity about the epic adventure they had west of the Mandan villages after he returned to St. Louis, their journals do not record it.

Instead, they note only that he reported a Mr. McClellan was a short distance below. This was Robert McClellan, with whom Lewis and Clark were acquainted from their earlier army service. Now engaged in the fur trade, he was headed north to the Yankton Sioux in a twelve-oared keelboat that was, by Ordway's account, "well loaded down" with merchandise from St. Louis. The trader "rejoiced to see us," Ordway continued, and "gave our officers wine and the party as much whiskey as we all could drink." While camped at the future site of St. Joseph, Lewis and Clark learned from McClellan that people "were concerned" about them after hearing rumors the entire party had been killed or were prisoners of the Spanish.[1]

The captains were glad to disprove those stories, but they had heard some disturbing news. Accompanying McClellan was Joseph Gravelines, an interpreter whom they had asked in 1804 to escort an Arikara chief on a visit to the president. They learned Gravelines now had the unhappy duty of informing the Arikaras that their chief had died while in Washington. Jefferson had instructed Gravelines to present gifts to the tribe and to introduce improved methods of agriculture. In addition, he was to make every effort on this trip to learn something regarding the fate of the expedition, since there had been no report in more than a year.

Traveling with Gravelines was Old Dorion, who two years before had gone with Lewis and Clark to Sioux country and was now charged with helping Gravelines move safely through Sioux territory with the gifts intended for the Arikaras. Along with being told to "make every enquirey" about the long-absent Corps of Discovery, Dorion had been asked to persuade no more than six principal Sioux chiefs to visit the city of Washington the following spring. On learning this, the captains made what Clark called "Some Small addition to his instructions" by increasing the number of chiefs to ten or twelve.

Two days after meeting McClellan's boat, the explorers came on another group of traders from St. Louis. These young men, traveling in three large boats, "received us with great friendship," Clark wrote, "and pressed on us Some whisky for our men, Bisquet, Pork and Onions. . . . We continued near 2 hours with those boats, makeing every enquirey into the state of our friends and Country." That night, he added, "our party received a dram and Sung Songs untill 11 o'Clock . . . in the greatest harmoney."

On September 15 the expedition's boats passed the mouth of the Kansas River and arrived at the future location of Kansas City, Missouri.

The captains had been impressed when they first saw the site, and now Clark described their second inspection. "Capt Lewis and my Self assended a hill which appeared to have a Commanding Situation for a fort, the Shore is bold and rocky imediately at the foot of the hill, from the top of the hill you have a perfect Command of the river, this hill fronts the Kanzas and has a view of the Missouri."

On the outward journey, the expedition had paused for two days at the confluence, but on this day the boats continued despite strong head winds to Hay Cabin Creek, later called the Little Blue River. The weather was "disagreeably" warm, and Clark wrote that "if it was not for the constant winds . . . we Should be almost Suficated Comeing out of a northern Country . . . in which we had been for nearly two years."

A few miles above the Grand River, two days later, the explorers met another Captain McClellan, this one a John McClellan, an acquaintance of Meriwether Lewis's. From the trading parties they met on the river, Lewis and Clark would have learned that Jefferson was still president, having been reelected in 1804. Alexander Hamilton, the nation's first Secretary of the Treasury, was dead, killed in a duel with Aaron Burr, a longtime political adversary. In St. Louis, political enemies were calling for the removal of General James Wilkinson, the appointed governor of the Louisiana Territory, who had become the focus of numerous suspicions and accusations. Anxious for a clearer picture of the "political State of our Country," the captains talked with McClellan from noon until almost midnight.

They were told that the president "had yet hopes" for their return, but they "had been long Since given . . . [up] by the people of the U S Generaly and almost forgotton." That word gave added incentive to reach St. Louis as soon as possible and assure family, friends, and the nation that they had indeed survived. Early the next morning the boats raced on.

Western Missouri had no shortage of game and fish. Near the future site of Independence the hunters shot a buck elk, described by Clark as "large and in fine order." It was the last of nearly four hundred elk killed during the expedition,[2] and it was the only one killed by the Corps of Discovery in Missouri, although Clark described the region as abounding in "Bear Deer & Elk."[3] The following evening another member of the party caught a catfish estimated at one hundred pounds. Deer were plentiful, and Ordway reported seeing a black bear running into a thicket. The animals were there, but the speed of the downriver journey and the party's lack of horses didn't give the hunters time to search out game and still keep up with the boats. Just below the Grand River, they told the captains they had not been able to kill anything. In place of meat, the expedition turned to pawpaws, a far more accessible food that had the added advantage of requiring no cooking or other preparation.

After two years of subsisting chiefly on deer and elk, meat that was usually tough, often dried, and sometimes "so much spoiled" that they ate it through "mear necessity," the men welcomed this change in diet. Ordway wrote that the pawpaws, "which our party are fond of . . . are a kind of fruit which abound in these bottoms and are now ripe." When an "emence Site of pappaws" was spotted, the men were even willing to brave "a number a rattle Snakes" to gather in the fruit.

On September 18, Captain Clark recorded in his journal that while the party was "entirely out of provisions [and] Subsisting on poppaws," the men "appear perfectly contented and tell us that they can live very well on the pappaws." His next observation in this same journal entry would seem to be totally unrelated: "One of our party J. Potts complains very much of one of his eyes which is burnt by the Sun from exposeing his face without a cover from the Sun. Shannon also complains of his face & eyes &c."

The following day, with the crews making good speed and stopping only long enough to gather more pawpaws, Clark added to his description of the problem:

A very singular disorder is takeing place amongst our party that of the Sore eyes. three of the party have their eyes inflamed and Sweled in Such a manner as to render them extreamly painfull, particularly when exposed to the light, the eye ball is much inflaimed and the lid appears burnt with the Sun, the cause of this complaint of the eye I can't [account?] for. from it's Sudden appearance I am willing to believe it may be owing to the reflection of the Sun on the water.

Clark's diagnosis doesn't seem to have convinced him, and it hasn't convinced later readers of the journals. Sun exposure seems unlikely as the principal cause, since the men had been exposed to the sun throughout much of their twenty-eight-month journey. Various ailments, including infectious conjunctivitis—pinkeye—have been suggested,[4] but a case can also be made for a link between the "singular disorder" he described and the fact that the men were "Subsisting on poppaws."

The pawpaw, *Asimina triloba*, is found in the central and southern United States. The tree grows in rich soil along streams, and its bright

Pawpaws were considered a delicacy by the Corps of Discovery, but the fruit may have caused the eye and skin discomforts Clark described in his journal. *Casey Galvin.*

green fruit, shaped like a stubby banana, matures in autumn to a brownish color. Although the flesh of the pawpaw is edible and considered by some to be tasty, susceptible individuals can experience an allergic reaction from eating or even touching the fruit. Such "poisoning by contact," according to one authority, can take the form of either mild dermatitis or "a painful irritation and inflammation."[5]

During the hottest and most humid week they had experienced that year, the men would have been frequently wiping sweat from their faces, transferring any allergens from their calloused and insensitive hands to the far more vulnerable area of their eyes and eyelids. Because sunlight is known to intensify the allergic reaction, Clark would be correct that the condition had something to do with the "reflection of the Sun on the water."

Below the mouth of the Osage River, on the westward journey through Missouri in 1804, Meriwether Lewis had collected a plant "known in Kentuckey and many other parts of this western country by the name of the yellow root." Better known today as goldenseal (*Hydrastis canadensis*), it was, he believed, "a speady and efficasious

Lewis believed "yellow root," known today as goldenseal, was an effective remedy for sore eyes. *Casey Galvin.*

remidy" for the "violent inflamation of the eyes" that was "a disorder common in this quarter."[6] Lewis wrote detailed instructions for preparing and using the root, which include placing pieces of carefully washed roots in a bottle or vial, adding water (preferably rain water), and applying the liquid to the eyes with "a piece of fine linin." His notation that it also "makes an excellent mouth water" accords with the more common use of goldenseal as a treatment for mouth sores, while applying it to the eyes seems almost as dangerous as bleeding a man suffering from sunstroke or giving harsh purgatives for undiagnosed abdominal pains. But Lewis's medical treatments were generally in line with the practice of the time, and his knowledge of herbal remedies was far above average.

Clark's journal makes no mention of whether goldenseal was tried or even remembered on the return voyage, but by the morning of

September 20, 1806, the eye problem had become so severe that three men were unable to row. The captains decided to leave behind a pair of twenty-eight-foot canoes Clark had ordered built on the Yellowstone two months earlier, and with the men and their supplies aboard the remaining boats, the party proceeded to the confluence of the Osage and Missouri Rivers. The expedition had spent two days there in June of 1804 while the captains took observations, but in 1806 the boats continued, passing the Gasconade River by noon.

While the rivers and streams Clark had mapped on the way west served as markers to tell the men they were now in the final miles of their journey, it was the sight of cows along the bank that "Caused a Shout to be raised for joy." Near sunset, after covering sixty-eight miles, the canoes arrived at the tiny French settlement of La Charette. Although it was the farthest outpost of white civilization on the Missouri River, to the members of the Corps of Discovery it was proof they were almost home.

The men fired three rounds as they drew near the shore and were answered by several trading boats, whose crews provided the explorers with a "very agreeable supper," one that was apparently a welcome change from the recent monotony of pawpaws. With rain imminent, the boatmen offered the captains a tent for the night, while residents of La Charette invited other members of the party to visit their simple cabins. Clark noted that villagers and traders alike were "astonished" at seeing the men and "informed us that we were Supposed to have been lost long Since." Expressions of wonder that they had survived and returned were repeated in the miles ahead. As the canoes moved downriver the next day, passing settlements that had not been established when the expedition began, the people at each "were Surprized to See us," Ordway wrote, "as they Said we had been given out for dead above a year ago."

About fifty miles beyond La Charette, the oarsmen quickened their pace as they drew within sight of St. Charles. Many of the town's residents were taking Sunday afternoon walks near the river, and a three-round salute from the boats brought a crowd to the banks. Here, as with all their encounters of the past week, delight was coupled with incredulity, for these people too "had heard and had believed," in Ordway's words, "that we were all dead and were forgotton."

That evening, with their men quartered in the town, the captains visited the homes of a few prominent citizens but had to decline several other invitations. "The inhabitants of this village," Clark wrote, "seem to vie with each other in their politeness to us all." The kindness of the St. Charles residents, coupled with a heavy rain, delayed departure until almost noon on Monday. With their safe return known and their men "all Sheltered in the houses of those hospitable people," the captains talked with their hosts and used some of the time to write letters. They could slow the pace in these last few miles, because in contrast to the ten weeks needed to cross Missouri in 1804, the return trip, with the current in their favor, had been made in only two.

Their next stop was Fort Belle Fontaine, on the south bank of the Missouri River at Cold Water Creek. Located about four miles west of the Mississippi, the fort had been built a year after the expedition's departure, but Lewis and Clark had learned of it during their conversations of the past week. The returning party was honored with a salute fired from the artillery company's field pieces and greeted by Colonel Thomas Hunt. Like Lewis and Clark, Hunt had lived the history of his time. He fought at Lexington and Concord, was wounded at Yorktown, served under General Anthony Wayne in the old Northwest Territory, and held a series of commands leading to his appointment to the newly established Fort Belle Fontaine.

Sergeant Ordway found it a "handsome place" and discovered "a number of these Soldiers are aquaintances of ours." Along with reunions, reminiscences, and the chance to hear more about events of the past two years, the captains had the opportunity to visit the post's store the next morning and outfit Chief Sheheke in preparation for the Mandan leader's introduction to St. Louis society.

On the final day of the voyage, the boats left the Missouri River and entered the Mississippi, stopping only briefly at the site of Camp Dubois, where the recruits had spent a winter preparing and waiting for the journey that was now drawing to a close. At noon on September 23, 1806, the canoes of the expedition arrived at the riverfront in St. Louis, where the captains allowed their men to fire three volleys "as a Salute to the Town." In return, they received a "harty welcom" from the large crowd of cheering St. Louisans who gathered at the water's edge when word came overland that the boats of the explorers would soon be in sight.

During the two years and four months of their journey, the men of the Corps of Discovery traveled over seven thousand miles and were eminently successful in carrying out the instructions of President Jefferson. They followed the Missouri River to its source, mapping that river and its tributaries. They traversed the Rocky Mountains, becoming the first citizens of the United States to cross the Continental Divide. They found the Columbia, the great river of the West, and followed it to the Pacific.

Lewis and Clark met the Indians of the Northern Plains and Northwest as official representatives of the United States and told the Indians the U.S. government wanted peaceful relations and trade. As they met tribes along the route, they noted their appearance, attire, and activities and systematically collected Indian vocabularies.

They sent artifacts to President Jefferson from Fort Mandan, and throughout the journey they collected animal and plant specimens, including more than two hundred plants Lewis dried and preserved.[7] They described species not known outside the West and helped define the range of known species. Their reports of vast numbers of beaver on the upper Missouri stimulated the developing American fur trade, which was centered in St. Louis.

Lewis's detailed scientific descriptions, combined with the daily accounts written by Clark, Ordway, and at least four other members of the expedition, constitute an extraordinary response to Jefferson's directive to keep multiple records of the voyage. In the words of one historian, the journalists of the Lewis and Clark Expedition became "the writingest explorers of their time."

And despite the difficulties and dangers they faced, their mission was accomplished with remarkable camaraderie and with the death of only one man. Two hundred years later, the Lewis and Clark Expedition remains an inspiring epic of diligence, courage, intelligence, and endurance.

# 5

## MISSOURI SEQUELS

M embers of the Lewis and Clark Expedition came from several parts of the nation. Of those who traveled to the Pacific, only George Drouillard, Pierre Cruzatte, and François Labiche had previously lived in Missouri. Following the explorers' return, many who made that epic journey would call Missouri home.

From St. Louis on September 23, Meriwether Lewis sent a long letter to President Jefferson beginning: "It is with pleasure that I anounce to you the safe arrival of myself and party at 12 OClk. today at this place with our papers and baggage." Continuing what he knew would be welcome news to the president, Lewis briefly outlined their successful route to the Pacific, noting that, despite difficulties, "navigation of the Missouri may be deemed safe and good." With an eye to the future fur trade, he informed Jefferson that the "Missouri and all it's branches from the Chyenne upwards abound more in beaver and Common Otter, than any other streams on earth." He then wrote that he would soon be coming to Washington, bringing a Mandan chief who had agreed to accompany him.[1]

Near the close of his letter, Lewis paid tribute to the man who had been his coleader throughout the hardships and triumphs of the long expedition: "With rispect to the exertions and services rendered by that esteemable man Capt. William Clark in the course of late voyage I cannot say too much; if sir any credit be due for the success of that arduous enterprize in which we have been mutually engaged, he is equally with myself entitled to your consideration and that of our common country."[2]

Both men were again welcomed by the Chouteau family, and when Lewis's letter had been dispatched, he and Clark turned their attention to the hospitality offered by their St. Louis friends. There was a dinner at Auguste Chouteau's home and a banquet and ball hosted by prominent St. Louisans. Between events the captains paid "visits of form, to the gentlemen of St. Louis," some of whom no doubt had been part of their send-off delegation two years earlier. One house they would not visit was the one in which they had been frequent guests during the months of preparation. Early in 1805, while they wintered at Fort Mandan, Pierre Chouteau's home, along with all his personal property and papers, had been destroyed by fire.[3]

In the first weeks after his return, Lewis discharged the enlisted men (who would each receive double pay and warrants for 320 acres of public land west of the Mississippi), sold at auction supplies and equipment left over from the expedition, and prepared for his visit to President Jefferson. After about a month in St. Louis, the captains set out for Washington. Their party included Chief Sheheke and his family, York, several former members of the Corps of Discovery, Pierre Chouteau, who was the Osage Indian agent, and a number of Osage chiefs.

William Clark paused at Louisville to spend some time with his family before going to Virginia, where he courted Julia Hancock. In mid-November the group divided again: Chouteau and the Osage Indians proceeded directly to the nation's capital, while Lewis and his party traveled by way of the Wilderness Road to Virginia. In his September 23 letter, Lewis had written that he was "very anxious to learn . . . whether my mother is yet living." Three months later he spent Christmas with her and also visited Monticello, where he saw displayed various artifacts he had sent to Jefferson.

Lewis's arrival in Washington with Chief Sheheke a few days after Christmas inaugurated a series of balls and banquets that included a White House reception hosted by President Jefferson. Washington society delighted in its Indian guests, and the president celebrated Lewis and Clark's fulfillment of his long-cherished dream.

Within weeks Jefferson sent to the Senate his nomination of Meriwether Lewis to be governor of Upper Louisiana. The nomination was approved on March 3, 1807. For reasons that remain unclear, Lewis waited a full year before returning to St. Louis to assume these duties. Some of that time was spent in Philadelphia, where he had gone to lay the groundwork for publication of the journals, but other factors were obviously at work in his procrastination.

In 1808, Lewis brought journalism to Missouri by persuading an Irish immigrant named Joseph Charless to come to St. Louis from Louisville to establish the city's first newspaper. Realizing Charless's press could have a major role in making citizens aware of new laws, the governor and others provided financial support, and the *Missouri Gazette* began publication in July of that year. The following year Charless published the *Laws of the Territory of Louisiana.*[4]

Governing the vast, unruly Louisiana Territory proved a formidable task. Lands previously occupied by Missouri's Indians were being encroached on by white settlers and also by members of eastern tribes who had crossed the Mississippi. As a result, the tribes fought each other and occasionally attacked settlers and traders. Disputes arose among white residents when their often casual land claims overlapped or when titles granted by the earlier French and Spanish governments were contested. Meanwhile, traders competed, despite the dangers, for permission to carry their goods to tribes along the Missouri.

Frederick Bates, a former Virginian whom Jefferson had appointed territorial secretary, was the acting governor during the year that Lewis delayed in returning to St. Louis after receiving his appointment. Confessing to being "totally in the dark" on issues such as granting licenses for trade on the Missouri, Bates sought the aid of William Clark, who had returned from Washington. When Lewis arrived, however, Bates greeted him with contempt.[5] The hard feelings may have originated years before, when Bates's father wanted one of his sons appointed secretary to the president and was disappointed when Jefferson instead chose Meriwether Lewis. Whatever the origin, Bates manifested a bitter antipathy toward the governor, undermining and insulting him whenever possible.[6] Lewis, who had known support and cooperation throughout the expedition, found his new situation dismaying.

Also plaguing him was the problem of returning Chief Sheheke to his tribe. The difficulty arose from the death in Washington of an Arikara chief whom the captains, early in their journey, sent to visit Jefferson. Lewis had learned of the chief's death when the Corps of Discovery was nearing St. Louis in 1806, and as governor he faced the implications of that event. The once-hospitable Arikaras had become revengeful, forcibly preventing an attempt to escort Sheheke to his Mandan village. In May 1807, fourteen soldiers under the command of Nathaniel Pryor, a former sergeant in the Corps of Discovery, set out from St. Louis with the chief and his family. Traveling with Pryor's expedition were more than thirty traders led by the eldest son of Pierre Chouteau, recently graduated from West Point. At the lower Arikara villages, Arikaras and Sioux attacked the party, killing three of Pryor's men and seriously wounding several others. Pryor gave up his attempt to return the chief and instead brought him back to Fort Belle Fontaine until a solution could be found. When the Mandan leader grew tired of the fort, Pierre Chouteau arranged quarters for him in St. Louis.[7]

Meriwether Lewis is believed to have sat for this crayon portrait by Charles St. Mémin during an 1807 visit to Philadelphia. *Missouri Historical Society, St. Louis.*

Meanwhile, Jefferson warned Lewis: "Nothing ought more to be avoided than embarking ourselves in a system of military coercion of the Indians."[8] The two-year impasse was finally broken when Lewis contracted with the Missouri Fur Company to form a militia large enough to assure safe passage through Arikara territory. Manuel Lisa had organized the trading company only months before, and its officers included William Clark and Lewis's younger brother, Reuben. The 125-man force led by Pierre Chouteau was successful in returning Sheheke to his people without incident, but bills for the mission that Lewis submitted to the federal government were rejected with a harsh rebuke by William Eustis, secretary of war in the new administration of President Madison. Jefferson, who Lewis could be certain would not have threatened him with personal liability for such expenses, had recently retired to Monticello.

In September 1809, Lewis left for the nation's capital, where he planned to defend his disputed drafts and protest what he considered an affront to his honor. In a letter Clark wrote to his oldest brother, Jonathan Clark, he expressed his confidence in Lewis but also a deep concern for his state of mind: "I do not beleve there was ever an honester man in Louisiana nor one who had pureor motives. . . . [I]f his mind had been at ease I should have parted Cherefully."[9]

Meriwether Lewis never returned to St. Louis. On October 11, while traveling through western Tennessee en route to Washington, he died of gunshot wounds. Clark would later tell Jonathan he learned that Lewis repeatedly imagined in the last miles of his journey "that he heard me Comeing . . . and Said that he was certain [I would] over take him, that I had herd of his Situation and would Come to his releaf."[10] Lewis was buried a few hundred yards from the cabin where he spent his final hours, far from family and friends. Most historians today believe Lewis himself fired the fatal shots. He was thirty-five years old.[11]

When Lewis made his westward crossing of the continent, he was accompanied by his dog, Seaman. The Newfoundland is last mentioned in the expedition's journals on July 15, 1806, reportedly howling from the torment of mosquitoes. Eleven days later, Lewis and the three men who accompanied him along the Marias had their encounter with the Blackfeet that left two Indians dead and Lewis's party fleeing on horseback to save their own lives. If Seaman were with them, he would have been unable to keep up with the horses, leaving Lewis the hard choice of shooting his dog or abandoning him in a hostile environment. But if Seaman had been with one of the boats rather than with Lewis, he might have survived, and research by historian James Holmberg recently offered hope that the dog did indeed return to St. Louis.

In 1814, Timothy Alden, a man with scholarly credentials, published a five-volume collection of inscriptions copied from headstones, monuments, and other sources. One of these inscriptions came from a most unlikely source: a dog's collar. The inscription as recorded by Alden read: "The greatest traveller of my species. My name is SEAMAN, the dog of captain Meriwether Lewis, whom I accompanied to the Pacifick ocean through the interior of the continent of North America." The likely time, perhaps the only time, for the inscription to have been made was after the expedition's return.[12]

The collar, with its amazing inscription, was in a small museum in Alexandria, Virginia. Many of the museum's artifacts were lost in a fire in 1871, and records of its holdings are incomplete, but Holmberg discovered that the museum wrote to William Clark on August 21, 1812, thanking him for the "Curiosities" he had presented. Of a "curiosity" believed to be part of Clark's donation, Alden wrote: "The foregoing was copied from the collar . . . which the late gov. Lewis's dog wore after his return from the western coast of America. . . . After the melancholy exit of gov. Lewis, his dog would not depart for a moment from his lifeless remains. . . . He refused to take every kind of food . . . and died with grief upon his master's grave."[13]

No witness to Lewis's last journey mentions a dog accompanying him, nor is there any ready explanation for how the collar would have come to William Clark if Seaman died in Tennessee unless, perhaps, it was removed from the dog when Lewis's possessions were being assembled after his death. In any case, Alden almost certainly recorded the story as it was related to him. If that was, in fact, Seaman's end, it would be in keeping with the service Lewis's dog had rendered and the esteem in which he was held.

Following the expedition, Missouri would become home to John Ordway. Born in New Hampshire about 1775, he had come west and was serving in Captain Russell Bissell's company of the First Infantry at Kaskaskia when Lewis recruited him for the expedition. Because Ordway was already a sergeant, his designation as one of the sergeants for the Corps of Discovery simply confirmed the rank he already held.

Intelligent, educated, and highly reliable, Ordway was given a series of assignments that began before the explorers left Wood River and continued after they returned to St. Louis. Ranking third behind Lewis and Clark in order of command, he had the unenviable task of taking charge of the recruits at Camp Dubois when both captains were absent. He also shared responsibility with the captains for keeping the roster and orderly book.

During the journey to the Pacific, he had such diverse duties as steering the keelboat, issuing provisions, and making guard assignments. In addition, he was sent on a number of special missions. When the Corps of Discovery divided into five sections on part of the return trip, Ordway led a group of nine men who traveled by river from the Three Forks to a rendezvous with another contingent at the Great Falls.

Lewis had instructed each of the sergeants to keep a journal, and in this, too, Ordway responded with intelligence and dependability. His clear, detailed, accurate record of the enterprise proved to be the only account to contain an entry for every one of the 863 days of the journey.[14] His journal alone would be reason to include him among the most valuable members of the Corps of Discovery.

Following the expedition's return to St. Louis, when Ordway had written "finis" on the last page of his notebook, there was still another assignment. He was selected to accompany Lewis and the Indian chiefs to the nation's capital for their meeting with President Jefferson. While in the East, Ordway probably spent some time with his family in New Hampshire.

By 1809 he had returned to Missouri, where he purchased a considerable amount of land on Tywappity Bottom, south of Cape Girardeau. His holdings included "two plantations under good cultivation [with] peach and apple orchards." The former New Englander who had traveled beyond the Rockies to the Pacific chose to live in Missouri, which he had described in 1804 as "the pleasantest place I have ever seen." A few years later, the region where Ordway settled became an unpleasant, even terrifying place.

Between December 1811 and March 1812, southeast Missouri experienced five earthquakes estimated at magnitudes of 8.0 to 8.8 on the Richter scale. Centered just southwest of New Madrid, in the bootheel of Missouri, they were felt from New Orleans to Ordway's native New Hampshire. With each quake the air filled with the smell of sulphur, the screaming of birds, thunderous roars, and dust that turned day into night. Most deaths occurred on the Mississippi, when boats were swamped or struck by collapsing banks and uprooted trees. Casualties among the settlers were apparently few. Their small frame houses typically suffered little more than a toppled chimney with the initial jolts and were unoccupied by the time subsequent tremors brought them down.[15]

Cracks in the earth changed the landscape, as did the thousands of trees that fell or died when their roots were torn. Some areas sank, forming new lakes, and other lands were flooded when rivers and streams changed their courses. Within the first year, aftershocks that could be felt numbered nearly two thousand. New settlement virtually stopped, and many established residents left. The region was described in 1819 as having "the most melancholy of all aspects of decay, the tokens of former cultivation and habitancy. . . . Large and beautiful orchards, left uninclosed, houses uninhabited . . . such was the face of the country."[16]

After 1811, Ordway's name disappeared from the public record until 1818, when his widow Elizabeth and her children applied for compensation from the federal government under a relief act for those who had suffered losses in the 1811–1812 earthquakes. In the document they were named as heirs of "John Ordway, Deceased, of the county of New Madrid." Although his peach and apple orchards apparently had not withstood the region's greatest natural disaster, Ordway had chosen to remain in Missouri.[17]

John Colter was not with the Corps of Discovery when the canoes arrived at St. Louis in September 1806, but he, too, became a Missourian after a series of adventures that made him one of the best-known members of the expedition. Colter, whose name would become part of the history of the American West, was born in Virginia about 1774. Some five years later he moved with his parents to Maysville, Kentucky, on the Ohio River, where he was recruited by Captain Lewis in 1803, becoming one of the "nine young men from Kentucky." A blue-eyed man of medium height and muscular build, Colter was endowed with quick intelligence, amazing stamina, and unfailing courage. During the winter at Camp Dubois, he demonstrated his skill as a hunter, and during the journey he was often given assignments in reconnaissance.[18]

When the returning explorers stopped at the Mandan villages in the summer of 1806, Colter asked permission to leave the expedition and join two trappers headed west. Clark wrote that both he and Captain

Lewis "were disposed to be of Service to any one of our party who had performed their duty as well as Colter had done."[19] He left with not only their blessings but with powder, lead, and other useful articles provided him by the captains and their men.

After trapping along the Yellowstone River, he was again headed down the Missouri when he met the party of Manuel Lisa, the St. Louis fur trader. Once more Colter reversed course. From a small trading post Lisa had established at the confluence of the Yellowstone and Bighorn Rivers, Colter set out alone, exploring the region at the base of the Teton Range in present-day Wyoming. Here he became the first white man to see the geysers and mudpots that characterize Yellowstone National Park. "Colter's Hell," a name first given to a neighboring area of thermal activity the explorer described, was later applied to Yellowstone Park.[20]

While trapping near the Three Forks in the autumn of 1808, Colter and another former member of the Corps of Discovery, John Potts, found themselves surrounded by hostile Blackfeet. Potts was killed as soon as he shot one of the Indians. Then, according to a story that has become legend, the Indians decided to have some sport with Colter before killing him. Stripped of his clothing, he was ordered to begin running. To the surprise of the Blackfeet, he outran all his pursuers but one, whom he killed with the warrior's own spear. This act of self-defense, according to biographer Burton Harris, was the only time Colter is known to have taken a human life. After plunging into a river, he remained with only his nostrils above water while the Indians spent several hours searching the riverbanks. When they finally left, Colter swam downstream a short distance and completed his escape by walking for eleven days—without clothing, moccasins, or food, except for roots and tree bark—until he reached the safety of Manuel Lisa's fort.[21]

In 1810, after other narrow escapes from death, he returned to St. Louis, where he met with William Clark and supplied him with valuable information for a map of the West that Clark was preparing for publication along with the journals of the expedition. Then, with his western adventures ended, Colter began a new life as a Missouri farmer. His land lay just across the river from La Charette, which had been the last white settlement the explorers saw when they set out in 1804. Some sixty miles west of St. Louis, where the Big and Little Boeuf Creeks flowed into the Missouri, he spent his few remaining years with his wife, whom he had married after his return, and their son, Hiram.

John Colter returned from his western adventures to become a farmer in this area near New Haven, Missouri. *Ann Rogers.*

In November 1813, at the age of thirty-eight, Colter died, apparently of jaundice. Most believe he was buried on a promontory overlooking the Missouri River at a place later known as Tunnel Hill. Colter's remains are thought to be above the old railroad tunnel or in the adjacent fill area. The exact location of his and other graves was seemingly lost when the Missouri Pacific Railroad cut a new passage through the hill in 1926.[22]

Although many of the men who traveled with Lewis and Clark lived in Missouri after the Corps of Discovery returned, George Drouillard had lived in Missouri before being recruited for the journey. Fathered by a French Canadian, Drouillard had been born in or near Canada, but he lived as a boy in the Cape Girardeau area with his Shawnee mother. In what would become southeast Missouri, he grew up not far from Louis Lorimier's trading post and home, which Captain Lewis visited as the explorers moved up the Mississippi in 1803.[23] At the time of Lewis's visit to the commandant and his Shawnee wife, Lewis wrote that he was bringing Lorimier letters of introduction, including one from "a Mr. Drewyer, a nephew of the Commandt's." It is possible that the nephew was in some way related to George Drouillard.

Cape Girardeau was the address Drouillard gave after the expedition when he purchased the land warrants of two of his fellow explorers, but he, like Colter, was drawn to the West and life beyond the frontier. In April 1807, less than seven months after the Corps of Discovery returned to St. Louis, Drouillard set out again on the Missouri River. This time he was in the company of Manuel Lisa, engaged in what has been called "the first organized trading and trapping expedition to ascend the Missouri to the Rocky Mountains."[24]

Among the fifty or sixty men traveling in a pair of keelboats were several other former members of the Lewis and Clark Expedition, whom Drouillard had recruited for this new venture. His presence and theirs no doubt had a role in Colter's decision to join Lisa's party when he met it at the mouth of the Platte. Colter, in turn, may have influenced Lisa in his decision to leave the Missouri at its junction with the Yellowstone and follow that river to the area where Colter had trapped the previous winter. Friendly Crow Indians willing to trade and an abundance of beaver made the fifteen-month trip a commercial success, but Drouillard learned on returning to St. Louis that he was to stand trial for murder.

The charge stemmed from an incident early in the 1807 journey. Near present-day Jefferson City, about 120 miles into the voyage, one man took some supplies and deserted. Lisa ordered Drouillard to bring him back "dead or alive." Shot by Drouillard, the man died from his wound while being taken by canoe to St. Charles. Public sympathy was with Drouillard and with the view that traders must be allowed to maintain discipline. The jurors concurred and took less than fifteen minutes to return a verdict of not guilty. Although Drouillard pointed out that he had been following Lisa's orders, he also expressed regret for his action. Meriwether Lewis had written following the Lewis and Clark Expedition that Drouillard had performed his duties "in good faith and with an ardor which deserves the highest commendation." Of the incident during Manuel Lisa's trading expedition, a biographer has noted: "Drouillard was ever faithful in the discharge of a duty. It was unfortunate that this time his zeal in this respect should have carried him too far."[25]

He again left St. Louis and returned to the West, where his skills as a hunter and interpreter had been so valuable to Lewis and Clark. In 1810, while trapping near the Three Forks, where Potts had died and Colter had made his famous escape, George Drouillard was killed by Blackfeet.

Another member of the Corps of Discovery who made Missouri his home was George Shannon, the youngest man recruited for the expedition. Born in Pennsylvania, he was about eighteen years old when he joined Meriwether Lewis for the trip down the Ohio River. Along with Clark's recruits, Shannon was one of the "nine young men from Kentucky."

The following summer the young hunter had caused concern when he became separated from the expedition as it moved through Sioux country early in the journey. A year later, near the Jefferson River in present-day Montana, Shannon was again hunting away from the main party. There was an unplanned change in the route, and Lewis's message regarding this disappeared when a beaver carried off the pole on which the note had been placed. But Shannon was able to determine which course the party had taken and rejoined it within a few days.

Writers who say he was "always getting lost" ignore the fact that twice in a twenty-eight-month journey is not "always," especially when the hunters had to range some distance in search of game and then estimate how far the boats had traveled during their absence. Historians generally acknowledge that he was one of the most intelligent members of the force, and the captains entrusted him with a number of missions.

These did not end when the explorers reached the St. Louis riverfront. One postexpedition assignment was the ill-fated attempt in 1807 to return Chief Sheheke to his tribe. In the attack that killed three of Pryor's men, Shannon received a bullet wound in the leg. Gangrene developed as he was being brought back to Missouri, and his leg had to be amputated. He remained for months in the hospital at Fort Belle Fontaine, the military post where, a year earlier, the Corps of Discovery had spent the final night of the expedition.[26]

In 1808, Shannon left the St. Louis area to begin two years of college studies at Transylvania University, in Lexington, Kentucky, and he remained away from Missouri for twenty years. Throughout those years he stayed in contact with William Clark, who continued to show concern for him. In 1811, Clark asked Shannon to go to Philadelphia to assist in the long-delayed publication of the expedition's journals. Nicholas Biddle, who was editing the manuscripts, reported back that his assistant was "very intelligent & sensible" and that Clark had made a wise decision in sending him.[27] During this period, Clark offered to help establish Shannon in trade at St. Louis, adding that if he preferred to pursue legal studies, he would have Clark's encouragement in that as well. Shannon chose the latter course, entered law practice, served in the Kentucky legislature, and became a judge.

Like others in the Corps of Discovery, he had received a land warrant and double pay for his service on the journey. Later he received compensation for the disability resulting from his service with Pryor, and, largely through Clark's efforts, this disability payment was increased.[28]

After an absence of two decades, he returned to Missouri, where some of his family was living, and at St. Charles, which he had first seen as an eighteen-year-old on an epic journey west, George Shannon became a U.S. district attorney for Missouri. In 1836, at the age of fifty-one, he died at Palmyra, Missouri, where he had gone for a trial. He was buried in the cemetery at Palmyra, but the exact location of the grave is not known.

In January 1837, the Missouri legislature passed an act stating that when a designated region of southeast Missouri had sufficient population to be organized into a county, it would "be called Shannon, in honor of the late Hon. George Shannon." Four years later Shannon County was named for this member of the Lewis and Clark Expedition.

Robert Frazer, along with at least nine other men in the Corps of Discovery, was a Virginian by birth, but he settled in Missouri following a postexpedition assignment. Like Ordway, he was part of the group accompanying Captain Lewis and Chief Sheheke to Washington. After festivities at the White House and celebrations throughout the capital, Frazer returned to the area that would be his home for the rest of his life.

In the late 1820s Clark wrote that Frazer was living on the Gasconade, the winding river that enters the Missouri west of Hermann. It was in nearby Franklin County that Frazer died in 1837, thirty-three years after he and his fellow explorers had camped at the mouth of the Gasconade on their outward journey.

Not long before his death, Frazer shared with neighbors his account of that great journey. John R. McBride noted in 1884: "He was a frequent visitor at my father's house in Franklin County, Missouri, and I can distinctly recall many of his conversations." Remembering him as "a man of education and talent," whose journal was "in many respects more interesting than that of his commanders," McBride wrote: "In his declining years, the delight of the old explorer was to sit by the fireside of some friend, read extracts from this journal written thirty years before, and add incidents from memory to the written tale."[29]

Frazer intended his account to have far wider circulation. Within a month after returning, he obtained Lewis's permission to have the journal published. A proposal for its publication, made in October 1806, announced that it would describe the rivers, mountains, Indian tribes, animals, and "a variety of Curious and interesting occurrences" in a work of about four hundred pages. The book would sell for three dollars and would be "put to the press so soon as there shall be a sufficient subscription to defray the expenses."[30]

He was the first to submit a proposal, but Sergeant Patrick Gass followed within a few months. Lewis, growing concerned that other journals would appear before his more complete and scientific account, then attempted to discredit them. A heavily edited version of Patrick Gass's journal appeared in 1807, but unfortunately the journal telling the story of the expedition as seen through the eyes of Private Frazer has never been located. However, Frazer's map of the region traversed by the Corps of Discovery is now in the Library of Congress and has been published in the *Atlas of the Lewis and Clark Expedition,* the first volume of Gary Moulton's edition of the journals.[31]

Frazer's journal may yet be found. Sergeant Ordway's detailed journal was found and published after being missing for a century. In 1953, Clark's account of the winter at Camp Dubois, the only known record of those five months, and his field notes written during the expedition's westward crossing of Missouri were discovered in a rolltop desk in an attic. A University of Illinois professor came upon a previously unknown version of Joseph Whitehouse's journal in 1966.[32] Perhaps a box of yellowed papers, an old desk, or a Franklin County attic holds the long-neglected journal of Robert Frazer.

During their journey through Missouri, the explorers made no mention of seeing Daniel Boone, but one member of the party is said to have spent time with this legendary figure after the expedition returned. John Shields, born in Virginia, was about fifteen when he moved with his family to Tennessee. Considered one of the "nine young men from Kentucky," Shields was among those Clark recruited in answer to Meriwether Lewis's instructions to assemble some young, unmarried men. Shields was thirty-four, by far the oldest of the recruits, and he was married, but Clark chose him because he knew this man would bring essential skills to the expedition. In addition to proving one of the best hunters on the journey, Shields was, according to one biographer, "the head blacksmith, gunsmith, boat builder and general repair man for anything needed."[33] When the Corps of Discovery returned, Lewis wrote that Shields's contributions, especially "the skill and ingenuity of this man as an artist, in repairing our guns," deserved special recognition.[34]

Not long after parting from the men of the expedition, Shields joined another man well acquainted with guns, hunting, and life in the wilderness. For a year he trapped in Missouri with Daniel Boone, his kinsman, who was living on the Missouri River about fifty miles west of St. Louis. The following year he trapped with Boone's brother, Squire Boone, in Indiana. In 1809, at the age of forty-one, John Shields died.[35] His life was not long, but within a span of five years it had been shared with Daniel Boone on the Missouri frontier and the men of the Lewis and Clark Expedition on their epic journey to the West.

A man who also served in the important roles of hunter, blacksmith, and gunsmith was William Bratton. Another of the "nine young men from Kentucky," he had moved from Virginia with his family in about 1790. He returned to Kentucky following the expedition but later came to Missouri and lived near John Ordway in the New Madrid area before enlisting for the War of 1812. The 1811–1812 earthquakes in southeast Missouri and their seemingly endless aftershocks may have been factors in Bratton's decision to settle in Indiana after the war and raise his family there. A monument in the pioneer cemetery at Waynetown, Indiana, notes that he "Went in 1804 with Lewis and Clark to the Rocky Mountains."[36]

In 1901, ninety-five years after Meriwether Lewis signed and dated Bratton's release from the Corps of Discovery, it was learned that one of his daughters, Ella Fields, who was then living in Chillicothe, Missouri, had preserved the original document. In the release, Lewis expressed "sincere thanks" and noted "the ample support which he gave me under every difficulty, the manly firmness which he evinced on every necessary occation, and the fortitude with which he boar the fatugues and painfull sufferings incident to that long Voyage."[37]

Like Bratton, George Gibson was another of the "nine young men from Kentucky." They, along with Joseph Field, formed the small group sent out soon after Christmas in 1805 to establish a salt-making camp on the Oregon Coast. It has been suggested that these three Kentuckians were given the assignment because high-quality salt was made at springs south of Louisville, and Clark was aware that one or more of them knew the process.

Salt making was not the only skill Gibson demonstrated while with the Corps of Discovery. Two years earlier, at Wood River, he was named the "best" in the first shooting competition between Clark's men and the local residents, and his ability as a hunter was noted repeatedly in the journal Clark kept throughout the winter at Camp Dubois. The following summer, as the explorers' boats moved up the Missouri, Gibson successfully brought one of the pirogues through a violent windstorm in northwest Missouri. Little is known about his few remaining years. He may have been wounded in the 1807 attempt to escort the Mandan chief back to his people. Records show that Gibson married after the expedition's return but died in St. Louis in 1809.[38]

Of the "nine young men from Kentucky," five became Missourians. Those who did not were Sergeant Floyd, who died during the expedition, Joseph Field, who died within a year of his return, his brother Reubin, who settled in Kentucky, and Nathaniel Pryor, who is buried in Pryor, Oklahoma, which was named for him.

When François Labiche traveled through Missouri with the Corps of Discovery, it was not his first voyage up the Missouri River. An experienced boatman and trader, Labiche knew the river, native dialects, and the tribes the explorers would meet in the first months of their journey. Like Private Pierre Cruzatte, he was an enlisted member of the permanent party, assigned to the keelboat at the start of the voyage. Later, in the crucial meetings with the Shoshones, Labiche, though unfamiliar with their language, served as interpreter. His knowledge of French and English made him a link in the long line of communication from the Shoshones to Sacagawea to the French-speaking Charbonneau to Labiche and finally to the captains. Lewis recommended that he receive a special bonus for his "very essential services" as an interpreter and asked him to serve again in this capacity when the captain took his Indian guests to Washington.[39] William Clark listed him as living in St. Louis during the 1820s, so he may be the "François Labuche" who, with his wife, Genevieve, had seven children baptized in St. Louis between 1811 and 1834.[40]

Alexander Willard, who was court-martialed early in the journey for falling asleep on guard duty, performed a trio of important roles for the Corps of Discovery as a hunter, blacksmith, and gunsmith. Living in Missouri after the expedition's return, he was employed by Lewis as a blacksmith in government service to the Sauk and Fox Indians. Clark's 1809 list of his employees in the Indian Department includes

Willard as a blacksmith "for the Shawnee." Long after the 1804–1806 journey to the Pacific, Willard continued to add interesting chapters to his biography. He became the father of twelve children, including sons he named Lewis and Clark, and he later traveled by covered wagon to California during the 1850s gold rush. He lived until 1865, into the age of photography, and appears with his wife, Eleanor, in a formal portrait. At the age of eighty-seven, Willard died in Sacramento.[41]

Richard Windsor was a member of the May 1804 court-martial that found Hall, Collins, and Werner guilty of being absent without leave prior to the expedition's departure from St. Charles. After the explorers returned, Windsor lived for a time in Missouri but then reenlisted in the army. In the 1820s, he was living along the Sangamon River in Illinois. Hugh Hall was apparently in the St. Louis area in 1809. Lewis's account book records a two-dollar loan to him in April of that year. John Collins, who, along with Hall, was court-martialed for getting drunk while the explorers were camped at the confluence of the Kansas and Missouri Rivers, may have lived in Missouri for a number of years after the expedition. Eventually he joined William Ashley's trapping party headed for the upper Missouri and was killed in 1823 by Arikaras. William Werner was another man whose transgressions early in the journey were apparently forgiven by the captains. After the expedition, according to one source, Werner "assisted General Clark for a time . . . in Missouri" when Clark was an Indian agent.[42]

After setting out from Louisville in 1803 with the captains and their recruits, York functioned not as Clark's personal servant but as a member of the expedition. The day after Christmas he and Joseph Whitehouse were "sawing with the whip Saws" at Wood River. On the journey west York killed an elk, brought down geese and other wild fowl, and shot buffalo on the plains.

Clark's journal singles him out as being especially attentive to the dying Sergeant Floyd, but York himself became the center of attention when the explorers met the Arikaras. "Those Indians wer much astonished," Clark wrote. "They never Saw a black man before, all flocked around him & examind him from top to toe." Sergeant Ordway, who often wrote in detail about the Indians, said of the Arikaras: "The Greatest Curiousity to them was York. . . . [A]ll the nation made a Great deal of him. the children would follow after him, & if he turned towards them they would run from him & hollow as if they were terreyfied, & afraid of him." York "Carried on the joke," not only demonstrating his physical strength but also telling the Indians that before Clark caught him he was "wild & lived upon people," finding "young children was verry good eating." Clark began to regret that York was making "him Self more turrible in thier view than I wished."[43]

The awe with which the Arikaras and other Indians regarded him and the easy acceptance the white recruits accorded him were only memories when the journey ended.[44] The return to his former status could be seen as reason enough for his discontent in the years that followed, but recently discovered letters show another cause. York had a wife in Louisville, whom he had married before the expedition, and he wanted to be with her. Instead, he was expected to live in St. Louis, Clark's new home.

In 1988, more than 175 years after they were written, forty-two letters from William Clark to his eldest brother, Jonathan, were found in the attic of a Louisville home belonging to Jonathan's descendants. Eleven of these letters mention York. On November 9, 1808, William told his brother that he was allowing York "to Stay a fiew weeks with his wife" but that "he wishes to Stay there altogether and hire himself." Although "he prefers being Sold to return[ing] here," Clark wrote, "I am determined not to Sell him."[45]

On his return to St. Louis, York did little to hide his resentment, and Clark wavered. In an apparent mixture of frustration and compassion, Clark did at some point hire out York to a man in Louisville. In May 1811, John O'Fallon, Clark's nephew, was in that city and reported to his uncle what he had learned about York's situation. He understood Clark's reason for sending York to Louisville was "that he might be with his wife," but O'Fallon had found her owner was preparing to move within a few months to Natchez, taking her along. Moreover, York's shabby appearance suggested he was not being treated well, and, the letter continued, he seemed "wretched under the fear that he has incurred your displeasure."[46]

Among the Clark letters is one written to Jonathan in 1805 from the expedition's winter quarters at Fort Mandan mentioning that York was sending a buffalo robe to his wife and another to a person identified as Ben. Ben was a former slave Clark had freed in 1802 "in consideration of the services already rendered to me" and because he believed "perpetual involuntary servitude to be contrary of the principles of natural Justice." These reasons should have compelled him to free York, and still he hesitated.[47]

Whether York returned to Missouri after 1811 or remained in Kentucky is unclear, but four years later he was the driver of a wagon, purchased by William Clark and a relative, for hauling in the Louisville area. Clark eventually gave him his freedom, providing him with six horses and a wagon for his own freight-hauling business between Kentucky and Tennessee. But notes made by the author Washington Irving, after a visit to William Clark in 1832, indicate York's business had failed and that he had died of cholera in Tennessee while trying to reach Clark in St. Louis.[48]

As Lewis and Clark entered the future state of Missouri on their return journey in 1806, they met a number of boats coming up the river. Along with other news from St. Louis, they heard that Zebulon Pike had recently set out from Fort Belle Fontaine to explore the sources of the Arkansas River. What the captains almost certainly did not learn then was that one member of Pike's expeditionary force had traveled two years earlier with the Corps of Discovery.

John Boley, after training at Camp Dubois, was passed over when the captains selected the men who would go to the Pacific. Instead, he was among those assigned to bring the keelboat back from Fort Mandan. But while he missed many of the dramatic chapters in the Lewis and Clark Expedition, he went on two explorations with Zebulon Pike. The first was Pike's search for the source of the Mississippi, a trip that began a few months after Boley returned from Fort Mandan in 1805. The next year, only weeks before the canoes of the Corps of Discovery swept through Missouri toward St. Louis, he again set out with Pike. After traveling up the Missouri River and then into present-day Kansas and Colorado, the party sighted the mountain that would bear Pike's name. Denied the chance to see the Rocky Mountains with Lewis and Clark, Boley saw the same mountain range hundreds of miles to the south.[49]

When he decided to settle down, he chose Missouri. His father, John Boley Sr., is said to have come to St. Louis from Pittsburgh in 1794. The younger Boley, one of six children, would have lived in the city prior to his service with Lewis and Clark. In 1823 he and his wife were living in Carondelet, once a separate community but now a part of St. Louis.[50]

Also returning from Fort Mandan in the spring of 1805 were about four French boatmen. Other engagés had returned the previous autumn. Because the expedition's journalists spelled names a variety of ways, especially the Frenchmen's names, it is difficult to identify all the boat-men or even know their precise number. Adding to the problem of identification, some used nicknames in place of their formal names.

The Chouteaus recruited the engagés from St. Louis and St. Charles, and what is known of them comes almost entirely from parish regis-ters in those cities. Peter Pinaut, whose mother was a Missouri Indian, was baptized in St. Louis in 1790; and Etienne Malboeuf, whose mother was also an Indian, was baptized in St. Charles in 1792. Jean Baptiste La Jeunesse married Malboeuf's sister in St. Louis in 1797. La Jeunesse and his wife had three children baptized in St. Charles. Paul Primeau was married in St. Louis in 1799, and Charles Hebert was married there in 1792. Some of Hebert's eleven children were baptized in St. Charles and others at Portage des Sioux, Missouri. Engagé Joseph Collin may have been the man of that name who was married at Portage des Sioux in 1818. The man in charge of the red pirogue and its crew of French boatmen was Jean Baptiste Deschamps. He may be the same man whose son, also named Jean Baptiste, was baptized in St. Charles in 1792.[51]

Fort Mandan was the place the French engagés and the half-dozen soldiers returning the keelboat parted from the expedition in 1805. It was also the place Toussaint Charbonneau, his young Shoshone wife, Sacagawea, and their infant son had joined the explorers. A multitude of paintings, statues, and writings portray Sacagawea as Lewis and Clark's indispensable guide on their transcontinental journey. That was not her role, but she did make significant contributions. To the Indians they encountered, she was a sign the explorers came in peace, because a war party would never be accompanied by a woman. As the Corps of Discovery neared the place where she had been kidnapped years earlier, her recognition of landmarks provided welcome assur-ance to the captains that they would soon find Shoshones. With her people located, Sacagawea acted as an interpreter, and her recognition of the chief as her brother aided in the negotiations. She again served as an interpreter when the explorers found members of her tribe else-where. Whenever possible, she gathered edible roots and berries that supplemented the party's meager rations or provided some healthful balance to a diet heavy in meat. On the return journey, when she was again in the area she knew from her childhood, she acted as a guide for Clark's segment of the divided party, directing the captain to use present-day Bozeman Pass as the best route to the Yellowstone River.

At the Mandan villages in the summer of 1806, Clark wrote: "We . . . took our leave of T. Chabono, his Snake Indian wife and their Son Child who had accompanied us on our rout to the pacific Ocean in the Capacity of interpreter and interpretes." It was not an easy parting for William Clark. On the day that the Corps of Discovery left the Mandans, he wrote in his journal that he had offered to take Charbonneau's son, "a butifull promising Child who is 19 months old to which they both himself & wife wer willing provided the Child had been weened." The couple believed, according to Clark, that in a year's time the boy would be old enough to leave his mother.

Three days later, while traveling downriver, Clark repeated his offer in a lengthy letter to Charbonneau, telling him that Sacagawea, who had shared "that long dangerous and fatigueing rout to the Pacific Ocian and back diserved a greater reward for her attention and services on that rout than we had in our power to give her at the Mandans." In gratitude to her and because of his affection for her son, he told Charbonneau that if he would bring the boy to St. Louis, "I will educate him and treat him as my own child." The letter presented Charbonneau with a variety of inducements, from promises of land and farm animals to offers of merchandise for trade with the Indians.[52]

The Charbonneau family did not arrive in St. Louis until December 1809. They may have wanted to make their trip earlier, but the hostile actions of Arikaras and Sioux against Sergeant Pryor's men affected river travel for two years. Three days after Christmas, the child Sacagawea carried on her back during the expedition to the Pacific was baptized at a small log structure that was the predecessor of the church St. Louisans know as the Old Cathedral. The baptismal record, in the parish's Register of Baptisms, was recently discovered by Bob Moore, the historian at the Jefferson National Expansion Memorial in St. Louis.[53]

Jean Baptiste Charbonneau, whom Clark called "my boy Pomp" and "my little danceing boy," is identified in the baptismal record not by name but by his birth date, February 11, the date on which Sacagawea, in the words of Sergeant Gass's journal, "made an addition to our number." The register shows that the godparents, Auguste Chouteau and his twelve-year-old daughter, Eulalie, signed their names, while Toussaint Charbonneau placed an X as his mark.[54] There is nothing to indicate whether Sacagawea watched as her son, by then almost five years old, received the Catholic Church's sacrament of baptism. Neither Lewis nor Clark witnessed the ceremony. Meriwether Lewis had died two months earlier, and William Clark was in Washington, D.C., dealing with some of the consequences of that death.

Because St. Louis had no resident priest in 1809, the baptism was performed by Father Urbain Guillet, a Trappist monk who crossed the Mississippi to spend Christmas week tending to the spiritual needs of the town. His order had recently established a small community of monks at Indian mounds located several miles south of those Clark explored during the winter at Camp Dubois. Monks Mound, now part of Cahokia Mounds State Historic Site in Illinois, takes its name from these nineteenth-century religious who lived at this site and farmed the mound's terraces.[55]

Records in St. Louis show that in 1810 Toussaint Charbonneau bought from Clark a tract of land on the Missouri River in St. Ferdinand Township, just west of St. Louis, but they also show he sold the land back to Clark the following year. The departure of Charbonneau and Sacagawea from Missouri in 1811 was described by Henry Brackenridge in his *Journal of a Voyage Up the River Missouri:*

> We had on board a Frenchman named Charboneu, with his wife, an Indian woman of the Snake nation, both of whom accompanied Lewis and Clark to the Pacific, and were of great service. The woman, a good creature, of a mild and gentle disposition, greatly attached to the whites, whose manners and dress she tries to imitate, but she had become sickly and longed to revisit her native country.[56]

In the winter of 1812, John Luttig, a fur company clerk from St. Louis, was working at Fort Manuel, located on the Missouri River just below the present border between North and South Dakota. On December 20 he made the following entry in his journal: "This evening the wife of Charbonneau, a Snake squaw, died of a putrid fever. She was a good and the best woman in the fort, aged about 25 years. She left a fine infant girl."[57]

The fort, established by Manuel Lisa and named for him, faced increasing threats by Indians and was abandoned in March 1813. Fifteen of Lisa's men were killed by Sioux as they came downriver,[58] but Luttig escaped and returned to St. Louis, bringing with him the baby girl he called Lizette. Because he stopped making daily journal entries in the final hours at the besieged post, there is no account of how he was able to care for an infant during the long trip from Fort Manuel. But a record in the Orphans' Court in St. Louis shows that he applied to be appointed guardian for Lizette and her brother, the child who had traveled with the Corps of Discovery. This record also shows Luttig's name was later crossed out and replaced with the name of William Clark.[59]

Charbonneau had been away from Fort Manuel when its occupants scattered. He continued to work as an interpreter and may have come to St. Louis sporadically, but there is no evidence he took any interest in his children. In 1839, when about eighty years old, he reappeared at the Office of Indian Affairs in St. Louis, asking to be paid for prior services as an interpreter at the Mandan villages, where a smallpox epidemic two years earlier had virtually eliminated the tribe. Joshua Pilcher, the superintendent, approved a modest payment, and Charbonneau once more went on his way.[60]

Jean Baptiste Charbonneau (sometimes called by his father's name, Toussaint) was educated in St. Louis, as Clark had promised, and traveled to Europe at the invitation of Prince Paul of Württemberg, whom he met on the Missouri River at the present site of Kansas City. After six years in Europe, Jean Baptiste returned to America to work as a trapper and as a guide for western travelers. Among those he escorted through the American wilderness was William Clark Kennerly, a nephew and namesake of William Clark. Later, Jean Baptiste tried his luck in the California gold rush and, after disappointments, was en route to goldfields in Montana when he died in 1866, at the age of sixty-one. His grave is in Danner, Oregon.[61]

In March 1807, when Meriwether Lewis was appointed governor of the Upper Louisiana Territory, William Clark was appointed by President Jefferson and confirmed by Congress as the territory's agent for Indian affairs, as well as a brigadier general in the territorial militia. Clark left the nation's capital, made a brief stop in Virginia, and then traveled to St. Louis to take up his new duties and begin what would be thirty-one years of residence in his adopted city.

On a second trip to Virginia, less than a year later, Clark married sixteen-year-old Julia Hancock. A few weeks after returning to St. Louis with his young wife, he set out on an almost three-hundred-mile trip to establish a fort and trading house near present-day Independence. From August 25 to September 22, 1808, Clark kept a journal of his overland trip to the site, the initial construction of the fort, and his treaty with the Osage Indians. Published in 1937 under the title *Westward with Dragoons*, the account bears many of the marks of his expedition journals. There is the same clear, terse style, the faithful recording of each day's events, and the abiding appreciation of the Missouri landscape.[62]

Leaving St. Charles, Clark rode at the head of a column of dragoons, mounted riflemen from the St. Charles area. Beside him rode their guide, twenty-six-year-old Nathan Boone, the youngest son of Daniel Boone. About the first twenty miles of the route followed the Boonslick Road, extending west from St. Charles towards Boone's salt-making operation. Just beyond the town the men moved through "Butifull . . . rolling Country intersperced with plains of high grass Most of them rich & fertile." Clark wrote of seeing turkey, partridge, grouse, and deer in the prairies, as well as elk near the headwaters of the Loutre River. He so admired the terrain near Bonne Femme Creek that his journal entry for August 31 contains four references to "delightful lands."

More than 170 miles west of St. Louis, the party "crossed a Cart Road leading from Boons lick," which was two miles to the north. Scouts sent by Clark found men working the salt lick Nathan Boone and his brother had established on land leased from James Mackay, whose knowledge of the lower Missouri had been of value to Lewis and Clark in 1804.

After camping for the night opposite present-day Arrow Rock, the troops crossed to the south side of the Missouri River at a place long used by Indians. Clark wrote that the spot provided "a fine landing on a Rocky Shore . . . and a gentle assent." They were assisted in the crossing by a pirogue left there by Captain Eli Clemson, who was bringing six keelboats filled with supplies and trade goods from Fort Belle Fontaine to the Fort Osage site. Also aboard were George Sibley, who would be in charge of the trading house, Reuben Lewis, younger brother of Meriwether Lewis, who would be sub-agent at the post,[63] and eighty-one soldiers under Clemson's command. One of those soldiers was named Joseph Whitehouse, and he was probably the same Joseph Whitehouse who kept a journal during the Lewis and Clark Expedition.[64]

After an overland journey of eleven days, Clark and his troops reached the Fire Prairie. When the War Department decided in 1808 to build a fort and trading house on the Osage River, both Lewis and Clark had written letters to Secretary Henry Dearborn suggesting a site higher up the Missouri and closer to the current Osage villages. Lewis specifically mentioned the Fire Prairie, and Dearborn accepted the recommendation.[65]

The day after his arrival at the Fort Osage site, Clark wrote: "Rose early [and] examined the Situation . . . found the River could be completely defended and Situation elegant, this Situation I had examined in the year 1804 and was delighted with it and am equally so now." Despite this statement, there is no specific mention of the site in either his 1804 field notes or journal, nor is there mention of it in his 1806 account of the expedition's return. However, it may have been noted by Clark on one of his Missouri maps that have been lost.

A detailed description of the Fort Osage site appears in the editions of the Lewis and Clark journals edited by Nicholas Biddle and later by Elliott Coues: "Directly opposite, on the south, is a high commanding position, more than 70 feet above high water mark, and overlooking the river, which is here of but little width. This spot has many advantages for a fort and trading-house with the Indians."[66] Reading the passage, one could easily assume Lewis or Clark had written it in 1804. In fact, the description is based on information Clark gave to Biddle in 1810, two years after Fort Osage was built.[67] A footnote in the Biddle and Coues editions mentions that the United States built a fort and trading house at this location. It is evident that Clark wanted the published account of the expedition to include a description of the site and a reference to the fort he had established.

On his first full day at the site in 1808, he "fixed on a spot for the fort and other buildings and ordered the Militia to clear the parade in front of their Camp." While Boone and an interpreter went to inform the Osage of their arrival, Clark ordered "a Strong Redoubt built on a Point to defend the Camp completely & placed a guard of 20 men in it." Two days later the foundation for the fort was laid. In his journal, Clark drew a preliminary sketch of the fort's design and provided a key to identifying its structures.

For twenty years Fort Osage was America's westernmost military outpost. It was also a government "factory," which meant the United States supplied trade goods in an effort to gain the Indians' favor and reduce British and Spanish influence over them. It was also hoped that these goods would lessen the Indians' dependence on vast hunting grounds, lands into which white settlers were moving, sometimes at the cost of their lives.

A week after Clark's arrival, Nathan Boone returned from the Osage villages with about seventy-five chiefs and warriors. Boone and the interpreter told Clark that in addition to these representatives, "all their villages were on their way." At the council held the next day, Clark told those present that attacks on whites must cease. The treaty he framed and the chiefs signed called for the Osages to relinquish their lands

In 1808, Clark traveled to a location near present-day Independence, where he supervised the construction of blockhouses, a factor's house, and other buildings. Today a reconstruction of Fort Osage stands on the site. *Ann Rogers.*

between the Arkansas River and a line running east from Fort Osage. The vast area included almost all of present-day Missouri south of the Missouri River. The benefits the Osages received were principally the government trading house and the promise of protection from feared eastern tribes. The chiefs expressed approval, cannon boomed, and the Indians celebrated.

Clark could not know then that the treaty would soon be rejected by other chiefs and replaced by one written by Meriwether Lewis. Language difficulties and misunderstandings were part of the problem, and apparently Pierre Chouteau, the Osage agent, urged the chiefs to reject any treaty that did not validate his land claims in the area—but neither he nor the Osages would gain by having the treaty rewritten. When the second version was presented by Chouteau in November 1808, the post was officially named Fort Osage. Some continued to use its earlier name, Fort Clark, and that is the name William Clark used when he placed it on his 1810 map of the West.[68]

By September 15, the fort Clark had sketched out only days earlier

Clark described the Fort Osage site as "elegant, commanding and healthy."
*Ann Rogers.*

had become a reality. A blacksmith shop was finished, the blockhouses were nearly complete, two houses for trade goods were under cover, and a road was being built. The dragoons who had come from St. Charles to offer protection and assistance were returning by land, but Clark, feeling "exceedingly unwell," boarded a boat for St. Louis. From the river he added to his journal one more description of the Fort Osage site: "We leave this Handsome Spot at 2 oClock and did not get out of Sight untill past 3 oCl[oc]k. The Situation is eligant Comdg [commanding] and helthy, the land about it fine."

The "factor," or man in charge of the trading post, was George Sibley, who had come to St. Louis from his native Massachusetts in 1805. When the War of 1812 left the fort undermanned and vulnerable to attacks by Indians, Sibley moved the post for a time to Arrow Rock, another site Clark had designated as a good location for a fort and a town. After Sibley left Indian service, he and his wife, Mary Easton Sibley, moved to St. Charles, where they endowed the school that has become Lindenwood College.

In addition to his government service as Indian agent, Clark joined a private, commercial venture as a founder of the Missouri Fur Company, established in 1809. Other officers included Manuel Lisa, Reuben Lewis, Pierre Chouteau, and Chouteau's son Auguste Pierre. It was the Missouri Fur Company that Lewis turned to after Nathaniel Pryor's unsuccessful attempt to return Chief Sheheke to the Mandans, and it was the federal government's questioning of Lewis's payment to the company that led indirectly to his death on the Natchez Trace.

Clark learned of that tragedy while he, his wife, and their infant son, Meriwether Lewis Clark, were on their way to Washington. The Clarks had left St. Louis at about the same time as Lewis but were traveling overland to visit family in Louisville. Lewis's intention was to go by boat to New Orleans and then up the coast, but hearing rumors of war and fearing his papers could fall into British hands, he decided to leave the river and continue by land. After seeing a newspaper report that Lewis had killed himself, Clark wrote to Jonathan, telling his brother: "I am at a loss. . . . His death is a turble Stroke to me, in every respect."[69]

For William Clark the untimely death of Meriwether Lewis not only caused deep personal sorrow but brought additional responsibilities. Three years after the explorers' return, the extraordinary journals that the two captains had kept during the expedition were still unpublished. After meeting with former President Jefferson, then in retirement at Monticello, Clark set about the tasks of assembling the notebooks, maps, and journals and of finding someone to prepare this material for publication. The Missouri Historical Society has notes made by Clark concerning this effort. One records his desire to find someone "to go to St. Louis with me to write the journal," and another reads: "Get some one to write the scientific part & natural history—Botany, Mineralogy & Zoology."[70]

From Monticello he went to Washington, where he was given the papers relevant to the expedition that Lewis had been carrying with him on his final journey. Then Clark went to Philadelphia, where a young lawyer named Nicholas Biddle agreed to prepare the manuscript. Although he proved capable, Biddle would not "go to St. Louis . . . to write the journal," nor would he "write the scientific part."

His location in Philadelphia meant Clark had to respond to numerous questions by mail, as well as those he answered at their initial meeting and during at least one other visit to the East. After a number of setbacks, including the bankruptcy of the intended publishing company, the first edition of the journals finally came off the press in 1814, but Biddle's lack of scientific knowledge led him to omit a large amount of valuable material. Only about fifteen hundred copies were made available, and two years after the publication Clark was still trying to obtain a copy for himself. Later he reported he was able to borrow one.[71]

In 1813 William Clark became the first governor of the newly formed Missouri Territory. Appointed by President Madison and later reappointed to three more terms, he served until Missouri attained statehood. Clark's tenure as governor, like Lewis's, came at a difficult time. The rapidly increasing white population disliked his order to respect Indian lands or face removal by the military. Land claims made during the earlier Spanish control of the territory were a source of continuing disputes, and tensions increased when the federal government moved eastern Indians west of the Mississippi, bringing them into conflict with both whites and established tribes. The War of 1812 pulled U.S. troops away from the frontier at the same time the British were inciting Indians against American settlers. Recurrent rumors that large numbers of Indians were massed for an attack on St. Louis sent waves of panic through the town. Meanwhile, small settlements in more remote areas were raided, and reports of Indian atrocities fueled a growing demand for revenge.

Throughout his eight years as territorial governor, Clark worked to secure the safety of Missouri citizens. One major effort was the council held at Portage des Sioux, in St. Charles County. For three months during the summer of 1815, Governor Clark, Auguste Chouteau, and Governor Edwards of Illinois met with chiefs from more than a dozen tribes. A number of peace treaties resulted, but for the many white settlers who wanted vengeance rather than mediation, Clark's approach to the problem was too conciliatory. Against this background he entered the race to become Missouri's first elected governor.

Clark was away from Missouri during much of the campaign. His twenty-eight-year-old wife had been in failing health and was staying at her former home in Fincastle, Virginia. Under criticism for spending too much time away from his official duties, Clark had just returned to St. Louis when he received word of Julia's death on June 27, 1820. Almost immediately he left again to make the long trip between Missouri and Virginia. Sorrow would be added to sorrow when his six-year-old daughter became ill and later died at the home of relatives in Kentucky.

By the time Clark and his sons returned to St. Louis, Missouri had its first elected governor: Alexander McNair. Clark had been defeated

This portrait of William Clark by Joseph Bush was painted in Louisville about 1817. *Filson Historical Society, Louisville, Kentucky.*

by a margin of more than two to one. He had not waged a vigorous campaign, but even if he had, it is unlikely the result would have been otherwise. His link to St. Louis's old order, the French with whom he had worked since he and Lewis first came to the town, estranged him from many of the new arrivals. Even more, Clark's belief that Indians and whites should peacefully share developing western lands was a policy out of favor on the American frontier of 1820.

In November 1821, William Clark married Julia's first cousin, Harriet Kennerly Radford, a widow with three young children. Missouri had entered the Union as the twenty-fourth state, and within a year Clark had a new appointment as Superintendent of Indian Affairs for the western region of the United States.

During and after his years as governor, Clark was involved in the educational and religious life of St. Louis. He was chairman of the first school board of trustees and was a founding member of the city's first Episcopal community, Christ Church, which later became Christ Church Cathedral. He encouraged Catholic priests and nuns to come to the area to educate both white and Indian children. Among those who responded to his invitation were Louis DuBourg and Joseph Rosati, who became bishops of the Louisiana Territory, and Peter John De Smet, a Jesuit who taught first at an Indian school in Florissant and then set out from St. Louis in 1840 to establish a mission for the Indians in the Bitterroot Valley. The native of Belgium had learned as a boy about the Flatheads and their homeland from Biddle's edition of the Lewis and Clark journals.[72] Philippine Duchesne, canonized by the Catholic Church in 1988 for her work as a missionary, came from France with other Religious of the Sacred Heart to establish schools for white and Indian children. Clark's stepdaughter would attend a school Philippine Duchesne founded in Florissant, a community just northwest of St. Louis. Describing an Easter Monday ceremony in 1822, she wrote: "We decorated the baptismal font and the children wore white dresses. I stood as godmother for the step-daughter of General Clark, the Governor, a very prominent man in St. Louis."[73]

(There is an earlier, coincidental link between Clark and Philippine Duchesne. In 1818, she and four other Religious of the Sacred Heart, newly arrived from France, were sent by Bishop DuBourg to St. Charles. There they established their residence and school in a house that a widow had previously divided into small rooms for boarders. The widow was Madame Duquette, and the house was the one in which Clark enjoyed a gracious dinner with the Duquettes while he was in

St. Charles just before the expedition set out from there in 1804.)[74]

Clark's public and private life found expression in the residence he built in 1816 at the corner of Main and Vine, on land now part of the Jefferson National Expansion Memorial. The two-story brick home was large enough for Clark's growing family, and in a wing at the south end of the mansion was his Council Chamber, a hundred-foot-long room displaying Indian artifacts, many of them presented by the Indians who visited him. Set out on tables and in cases, attached to the walls, and suspended from the ceiling were canoes, snowshoes, papoose cradles, beaded clothing, fur robes, feathered headdresses, and bear-claw necklaces. There were tomahawks, battleaxes, and bows and arrows, as well as peace pipes, cooking utensils, and musical instruments. And because the display was open to the public on request, William Clark's Council Chamber was the first museum west of the Mississippi.[75]

Months before Clark died, Meriwether Lewis Clark gave the natural history portion of his father's museum to the Western Academy of Natural Sciences in St. Louis, and a year earlier Albert Koch had procured many of the Indian pieces for his St. Louis Museum. Nothing from the original collection can be traced. The Indian portraits that visitors mentioned seeing on the walls of the Council Chamber disappeared and were not mentioned when Clark's estate was inventoried.[76] St. Louis lost a treasure when Clark's museum was dismantled.

One famous visitor to the Council Chamber was the Marquis de Lafayette, who years earlier had come from France to help America in its war for independence. A welcomed guest on return visits, he came to St. Louis in 1825 as part of an extensive tour of the United States. Clark received from Lafayette a mahogany camp chest, carefully fitted with a coffeepot, dishes, candlesticks, and other items useful for travel. (The Missouri Historical Society now owns and sometimes displays the chest.)

A frequent caller at William Clark's home was a Pennsylvania-born artist, George Catlin. At Clark's invitation, he painted portraits of many native chiefs as they met in the Council Chamber with the man they called the Red Head. Fascinated by the Indians he met and by Clark's descriptions of the Northwest, Catlin became the first artist to travel up the Missouri River, adding to his collection of Indian portraits and recording on canvas the Nebraska bluffs, Sergeant Floyd's gravesite, the Mandan villages, and other landmarks of the Lewis and Clark Expedition.

The individuals who had shared in that journey were not forgotten by William Clark. Sometime between 1825 and 1828 he used the cover of one of his account books to list the members of the expedition. After

George Catlin painted portraits of No Horns on His Head and Rabbit-Skin Leggings, two of the Nez Percé who visited Clark in 1831. *Smithsonian American Art Museum, Gift of Mrs. Joseph Harrison, Jr.*

most of the names he added a brief comment. Of the thirty-four listed, twelve are followed by the notation "dead." These include the names of Meriwether Lewis, John Ordway, John Colter, and Sacagawea. Six more names, including that of George Drouillard, are followed by the notation "killed."[77]

Among Clark's fellow Missourians in 1825 were François Labiche, Alexander Willard, and Robert Frazer. Other former members of the Corps of Discovery were living in Virginia, Ohio, Illinois, Kentucky, and Arkansas. Five names were listed without comment. Apparently Clark knew nothing of the fate of these men.

With many members of the expedition dead and the rest widely dispersed, Clark's attention in his later years turned increasingly to the Indians who visited his Council Chamber. In the autumn of 1831, a delegation of four Indians arrived in St. Louis to see Clark. One, called Man of the Morning, was from the Flathead tribe, and three were from the Nez Percé. They explained that they had been told of a Book of Heaven that gave power to white men. A council had decided the "Red Head" who had visited their tribes would tell them the truth and help in their quest.

An immediate concern was the health of the four Indians. They had traveled from west of the Rocky Mountains, making much of the trip in the heat and humidity of late summer. Already ill when they arrived in St. Louis, they found the climate, food, and customs totally unfamiliar. Two of the Indians died within a few days of each other. Both were baptized shortly before their deaths, and both were buried in the Catholic cemetery.

The following spring Clark booked passage for the two surviving Indians on the *Yellow Stone,* the first steamboat to ascend the upper Missouri River. George Catlin, who painted their portraits, was also on board for the historic trip to Fort Union, about 2,100 miles upriver from St. Louis. Although Clark had done what he could to assure the young men's safe return, he later received word that one of the Indians had died near the mouth of the Yellowstone River and that the last of the four, Rabbit-Skin Leggings, had been murdered by Blackfeet as he was nearing the end of his long journey home.[78] They had left St. Louis without the missionaries they hoped would accompany them, but within a few years missionaries came to their lands, most notably Father De Smet, known to the Indians as Black Robe.

The fact that Man of the Morning and his companions traveled so far to seek Clark's advice is a testament to the deep impression Lewis and Clark made on the Indians they encountered. Historian Bernard DeVoto observes that the loyalty to white Americans that these two explorers inspired among the Flatheads and Nez Percés endured through decades of injustice by other white men. Only when the outrage became unbearable in the 1870s did Chief Joseph rebel and lead a group of Nez Percés in a fighting withdrawal toward Canada. Until that time, DeVoto points out, "they not only had never attacked a white man, they had never been offensive to one."[79]

William Clark's reputation for truth and justice extended the length of the Missouri River and beyond. When Indians arrived in St. Louis, he was the person they sought out. He was, of course, the superintendent of Indian affairs, but more than that, they knew he would treat them with dignity, honesty, and fairness. He could not, as they soon learned, do everything, but what he could do for them he would. The white man's treatment of Native Americans is not a proud chapter in the nation's story, but Clark's role is, in DeVoto's words, "a bright strand in a dark history."[80]

On the evening of September 1, 1838, William Clark died at the St. Louis home of his eldest son, Meriwether Lewis Clark. The funeral two days later was carried out with full military honors. A company of soldiers escorted the hearse, which was drawn by four white horses. Clark's own horse followed, riderless and with boots reversed in the stirrups, symbolic of a lost leader. A mile-long cortege, said to have been the largest in St. Louis history to that time, accompanied the body to its resting place on the farm of Colonel John O'Fallon, Clark's nephew.

His body was later moved to Bellefontaine Cemetery, where a granite obelisk and a bronze bust of Clark now mark the site. At the base of the monument these words are inscribed: "William Clark, Born in Virginia August 1, 1770; Entered into Life Eternal September 1, 1838; Soldier, Explorer, Statesman and Patriot, His Life Is Written In the History of His Country."

The monument to William Clark at Bellefontaine Cemetery was dedicated during the St. Louis World's Fair in 1904, the year that marked the centennial of the start of the Lewis and Clark Expedition. *Ann Rogers.*

# 6

# LEWIS AND CLARK AND THE 1904 LOUISIANA PURCHASE EXPOSITION

B etween April 30 and December 1, 1904, nearly twenty million people answered the invitation to "meet me at the Fair." The fair was the St. Louis World's Fair or, more officially, the Louisiana Purchase Exposition. Over a thousand acres of Forest Park were turned into fairgrounds, taking advantage of the wooded terrain, hills, and lakes. Ornate exhibition palaces averaging seven acres in size appeared as an ivory city by day and, in the gaslight era, were even more magical at night when outlined by thousands of Edison's electric bulbs. Among the countless entertainments were a twenty-five-story "observation wheel" designed by a Chicago engineer named Ferris, band concerts directed by John Philip Sousa, the first Olympic games held on U.S. soil, and daily reenactments of the Boer War.

If anything so diverse could have a focal point, it was perhaps the Louisiana Purchase monument, designed by Karl Bitter, chief of sculpture for the exposition. The hundred-foot column rising from the edge of the Grand Basin had at its base allegorical representations of the Mississippi and Missouri Rivers. On one side of the column was a high-relief tablet depicting Monroe, Livingston, and Marbois at the Paris signing of the Louisiana Purchase.

About one thousand sculptures were approved for the grounds and exteriors of buildings. Like the palaces, almost all the statues were made of staff, a plaster-of-paris mixture not intended to be permanent. The exhibition buildings were ornamented with muses, griffins, and various allegorical figures, but in keeping with the exposition's theme, the portrait statues on the grounds related to the Louisiana Territory. Among those represented were Pierre Laclède, Marquette, Joliet, and Daniel Boone.

In the group of portrait statues designated "Pathfinders of the Purchase," a heroic (or larger-than-life) representation by Charles Lopez showed Meriwether Lewis in buckskin and moccasins. An account

The statues of Meriwether Lewis and William Clark at the 1904
World's Fair in St. Louis were the first to commemorate the leaders
of the 1804–1806 expedition. *Missouri Historical Society, St. Louis.*

from 1904 points out the symbols of Lewis's dual role on the expedition: "One hand grasps the long rifle and the other the commission
and papers of official character. . . . He was at once the official representative of the Jefferson administration and the head of a hardy
adventurous band finding a path to the Pacific."[1] The statue, which
stood in a prominent location along the Grand Basin near the Cascades,
was the nation's first to honor Meriwether Lewis.

Also along the Grand Basin and among the "Pathfinders of the
Purchase" was the first statue of William Clark. Sculptor F. W. Ruckstahl
used facial features from Charles Willson Peale's portrait of Clark and

showed the expedition's coleader in frontier garb and with powder horn and hunting bag.[2] He was shown carrying a rifle in his right hand, and on his left side he had his sword. The expedition's journals record that Clark drew his sword when threatened by Sioux early in the journey but gave the weapon to a Walla Walla chief during the return trip in exchange for an "eligant white horse" and other assistance provided to the explorers.

At the close of the fair, the statues of the captains were shipped to Portland, Oregon, and displayed at the 1905 Lewis and Clark Centennial Exposition. Then they disappeared.[3] When their real-life counterparts were nearing St. Louis in 1806, Clark had written that the residents were astonished to see the men thought "to have been lost long Since." Lewis and Clark had returned from the Pacific; their statues, making the journey a century later, were the ones to have been lost.

In what was described as "a gem of a garden" stood a statue of Sacagawea by Bruno Louis Zimm. This New York sculptor's creation was said to have evolved from his study of the Shoshone tribe and his use of a Shoshone woman as the model. Karl Bitter called it "a small monument of special interest . . . erected to the memory of the Indian woman who rendered such splendid services in connection with the Clark-Lewis expedition."[4]

The statue stood on a base similar to those of the Lewis and Clark statues but could have been considered small in comparison to them and other "heroic" monuments. With her passive expression and hands clasping a walking stick, she recalls Clark's description of the woman who endured "that long dangerous and fatigueing rout to the Pacific Ocian."

When the fair ended, there was no need to send the statue to the centennial in Oregon because a bronze representation of Sacagawea would be unveiled in Portland as part of that event. The staff figure made for the World's Fair in St. Louis was probably destroyed along with countless other objects when the grounds were cleared. Yet, Bruno Zimm's creation was not only the nation's first statue of Sacagawea but also the exposition's only figure among the nearly one thousand sculptures on the fair's buildings and grounds to depict a historical woman.[5]

Almost every state in the Union, as well as the Indian Territory, had a building for hospitality and special displays. The exteriors of many replicated historical structures: Tennessee reproduced Andrew Jackson's Hermitage; Mississippi replicated Beauvoir, the last home of Jefferson Davis; and California re-created the Santa Barbara Mission.

The statue of Sacagawea, by Bruno Zimm, was the only one on the fairgrounds to honor a historic woman. *Missouri Historical Society, St. Louis.*

To represent their state and honor the man considered the "patron saint of the exposition," Virginians looked to examples of Jefferson's architecture. "The choice lay between one of the university buildings designed by him and the home he designed for himself and in which he lived and died."[6] A replica of either would have been a fitting tribute, but Monticello, already standing at the time of the Louisiana Purchase, has strong links to that event and to the Lewis and Clark Expedition. It was at Monticello that Jefferson and Lewis made plans for the great journey. It was there that Jefferson displayed Indian artifacts sent back from Fort Mandan; and it was there that he met with William Clark to discuss publication of the journals that described the Louisiana Territory and recorded the explorers' achievements.

Photographs of the full-size replica of Monticello show an exterior faithful to Jefferson's design. Inside were displays of letters and other

The Virginia State Building at the World's Fair was a faithful replica of
Monticello. *Missouri Historical Society, St. Louis.*

historical documents. In one room the University of Virginia exhibited
its full-length marble statue of Jefferson by Alexander Galt (now in the
university's Rotunda building) and its Thomas Sully portrait of Jefferson
at seventy-eight in a fur-collared cloak.[7]

Jefferson's original tombstone, a six-foot-high obelisk designed by
him, was sent to the fair by the University of Missouri, whose Columbia
campus has been its home since 1885. (A professor at that university,
who was also a graduate of the University of Virginia, learned that the
original monument was being replaced, and he won it for Missouri
with the argument that it should be at the first public university
established in the Louisiana Territory.) The monument that originally
marked Jefferson's grave stood among the exhibits in the fair's Palace
of Education.[8]

September 23, the anniversary of the Corps of Discovery's 1806
return to St. Louis, was celebrated by fairgoers as Lewis and Clark Day.
William Weil's band played his "Lewis and Clark March," and a liter-
ary program was presented at the Oregon Building, a representation of
Fort Clatsop. While this structure bore little resemblance to the replica

The Oregon State Building, a representation of Fort Clatsop, was moved to the St. Louis suburb of Kirkwood when the fair closed. *Missouri Historical Society, St. Louis.*

constructed half a century later at Astoria, its shortcomings are understandable. Virginia could see Monticello, then privately owned, while Oregon could do little more than guess. Nothing remained of the original structure when the Oregon Historical Society purchased the Fort Clatsop site in 1901, and Clark's sketch showing two rows of rooms facing an inner yard was etched on the elkskin cover of a journal his granddaughter, Julia Clark Voorhis, had not yet released for publication.[9]

Oregon timber was brought to St. Louis for the structure, and dry moss was used to fill cracks between the horizontal logs of the main building. The replica included two blockhouses, apparently not part of the 1805 design, as well as logs set on end to form an eleven-foot-high stockade. One of the exhibits in the Oregon Building was a relief map showing the route of the expedition from the St. Louis area to the original Fort Clatsop.[10]

When the fair closed in December 1904, the huge exhibition palaces were quickly turned into mountains of crushed staff, usable only as landfill, while many of the state buildings, made of more permanent materials, were sold and moved. Some traveled as far as Iowa, Oklahoma, and New Mexico. At least seven stayed in the St. Louis area, and four of these remain standing. The Oregon Building was purchased by Anderson Gratz and moved to his property in the St. Louis suburb of Kirkwood. Rebuilt and renamed the "Wickiup," it was used by his daughter for "parties and other entertainments" before being destroyed by fire in 1913.[11]

By then its significance had been lost, but during the summer of 1904 its meaning was never obscured. A sign at the entrance told visitors: "This structure is a replica of Old Fort Clatsop, the winter quarters, 1805–6, of Captains Lewis and Clark with their company after they had, in the greatest of American explorations, crossed the continent to the Pacific."[12]

As part of the 1904 commemoration, a granite obelisk and bronze bust of William Clark were dedicated on October 2, at Bellefontaine Cemetery. Clark's body had been moved from his nephew's farm in 1860 and reinterred along with those of five family members at the cemetery north of the city. At the 1904 dedication, Mayor Wells spoke of Clark's untiring energy and the great good he had done for St. Louis; and David R. Francis, president of the Louisiana Purchase Exposition, said people in this part of the country were only beginning to appreciate Clark's achievements. Just before the unveiling of the monument, a Creek chief from the Indian Territory praised Clark as "a brave man and a man of mercy," who throughout a difficult mission "made friends with the Indians."[13]

A photograph from the next day's edition of the *St. Louis Republic* shows Clark family members posed on the obelisk's terraced base. Nine children seated on the steps represent what was then the youngest generation. Standing behind them is Mrs. Jefferson K. Clark, the widow of Clark's last son, whose will provided for his father's monument. Also pictured are John O'Fallon Clark, who had loaned the fair a portrait of his grandfather painted by Chester Harding; Julia Clark Voorhis; and her daughter, Eleanor Glasgow Voorhis.[14] The Missouri Historical Society would eventually receive from Julia Voorhis the Lewis and Clark journals she had inherited, including Clark's field book with his sketch of Fort Clatsop on the elkskin cover.[15]

David R. Francis and others who had worked to make the Louisiana Purchase Exposition a success proposed a memorial to Thomas Jefferson as a legacy of the fair. The exposition company and the City of St. Louis agreed their share of the profits should go to this purpose, and Francis persuaded Congress to contribute the federal government's share as well. Dedicated in 1913, the building is on the site of the fair's main entrance and features in its loggia a marble statue of Jefferson designed by Karl Bitter and a casting of Bitter's bas relief entitled *The Signing of the Louisiana Purchase Treaty.*

The structure was intended, Francis wrote, to "house our historical literature and official records and relics" in "worthy surroundings" where "they might always be accessible."[16] For this reason the Jefferson Memorial became home to the Missouri Historical Society, whose massive collection includes Lewis's telescope and watch, Clark's compass and magnet, a portrait of Lewis by Charles St. Mémin, numerous documents and correspondence related to the explorers who became governors, and, as the centerpiece of the collection, the journals of the Lewis and Clark Expedition, including Clark's elkskin-bound field book.[17]

# 7

## THE TRAIL TODAY

T he centerpiece of the Jefferson National Expansion Memorial is the 630-foot Gateway Arch, a symbol of St. Louis's role in the westward movement. Beneath the Arch is the Museum of Westward Expansion. David Muench's photographs form a large, colorful mural of scenes from the Lewis and Clark route. The explorers, Indians, and homesteaders are recalled in a series of exhibits, and there is a large display of Indian peace medals. Visitors can also view documentary films and ride the tram to the top of the Arch.

The church St. Louisans call the Old Cathedral, built in 1831, is at 209 Walnut, adjacent to the Arch grounds. At an earlier church on this site, Sacagawea's son was baptized in 1809. That log structure was also the church in which three of William Clark's children were baptized in 1814.

Across from the Arch is the Old Courthouse, another part of the Jefferson National Expansion Memorial, with exhibits related to early St. Louis.

A bronze tablet displayed in the lobby of St. Louis Place, an office building at 200 North Broadway, between Olive and Pine Streets, identifies the site as the place William Clark died. In 1838 the home of his son Meriwether Lewis Clark stood here. The plaque, when originally unveiled in 1906, was on the exterior wall of a bank then occupying the site.

William Clark is buried at Bellefontaine Cemetery north of downtown. From the Arch, take I-70 west to the West Florissant exit and go right (northwest) one mile. The cemetery office has maps and a list of famous Missourians who are buried at Bellefontaine.

The History Museum and Missouri Historical Society are housed in the Jefferson Memorial, located in Forest Park, west of downtown St. Louis. A fountain near the front entrance commemorates the Lewis and Clark Expedition. In the plaza, a large map etched in granite shows

The Gateway Arch on the riverfront symbolizes St. Louis's role as the gateway to the West. *Ann Rogers.*

the explorers' route and contains excerpts from the journals. In the loggia are Karl Bitter's statue of Jefferson and a bas relief of the signing of the Louisiana Purchase treaty. The museum offers exhibits and programs relating to early St. Louis, and those interested in Lewis and Clark may be fortunate enough to visit when items such as Lewis's telescope or Clark's elkskin journal are on display. The Historical Society's library on Skinker Boulevard, a few minutes' drive west of the museum, is a favorite place for scholars researching western history.

Powder Valley Conservation Nature Center, 11715 Cragwold Road, in the suburb of Kirkwood, features a half-scale replica of Lewis and Clark's keelboat, natural history exhibits and programs, and hiking trails. The lobby has a windowed viewing area for watching songbirds, wild turkey, and small mammals feeding. A three-thousand-gallon simulated pool with catfish, bass, and bluegill provides aquarium-style viewing on the lower level but from the upper level looks like a natural pool. Powder Valley is a favorite with families because it offers activities and exhibits that are interesting to both children and adults.

The Mercantile Library, on the St. Louis campus of the University of Missouri, at 8001 Natural Bridge Road in the suburb of Normandy, specializes in the history of St. Louis and America's westward expansion. The library includes a gallery of western art, featuring Chester Harding's full-length portrait of William Clark and Indian portraits by George Catlin, who painted in Clark's Council Chamber.

The stately General Bissell House, 10225 Bellefontaine Road, dates to about 1812 and was the home of the general who commanded nearby Fort Bellefontaine. When a captain, he was the commander at Fort Massac when Lewis and Clark stopped there in 1803.

Jefferson Barracks, overlooking the Mississippi in Lemay, about ten miles south of the Arch, replaced Fort Bellefontaine in 1826, with William Clark helping to choose the site. Exhibits tell the history of the post where Grant, Lee, Sherman, Longstreet, and Fremont were stationed. One diorama shows the Sauk chief Black Hawk guarded by a recent West Point graduate, Lieutenant Jefferson Davis. Clark visited here in 1833 at the request of Sauk and Fox Indians who wanted him to intercede on behalf of the chief who fought to reclaim Indian lands east of the Mississippi.

The St. Louis Zoo, located in Forest Park, is one of the best zoos in the United States, and admission is free (except for a parking charge). It offers a chance to see some of the species Lewis and Clark described.

A visit to St. Louis should include the Missouri Botanical Garden, 4344 Shaw Avenue, in south St. Louis. There is a large variety of indoor and outdoor gardens, as well as pools, fountains, a maze, a pleasant restaurant, and a bookstore and gift shop. Visitors can see beautiful specimens of many of Missouri's indigenous plant species that Lewis and Clark noted, including Osage orange and sycamore trees and a representative planting of prairie grasses.

The expedition's journals abound in favorable references to Missouri's prairies, which once covered 40 percent of the state. The Shaw Nature Reserve (formerly the Shaw Arboretum), thirty miles west of St. Louis at Gray Summit, I-44 exit 253, includes a seventy-eight-acre tallgrass prairie accented by native wildflowers. An observation deck provides an overview, while miles of hiking trails introduce visitors to the plants, flowers, birds, animals, and trees of the twenty-five-thousand-acre nature preserve. In addition to this extension of the Missouri Botanical

The sixty-acre prairie at Shaw Nature Reserve, formerly Shaw Arboretum, represents the vast grasslands that greeted Lewis and Clark as they traveled across the continent. As the season progresses, over seventy wildflower species bloom and the grasses grow up to ten feet into the sky. *Ann Rogers.*

Garden, a number of other prairies across the state have been preserved or restored through the efforts of the Missouri Department of Conservation, the Missouri Prairie Foundation, and the Missouri Department of Natural Resources.

The Missouri Department of Conservation and other organizations are working to provide a good view of the confluence of the Missouri and Mississippi Rivers.

### St. Charles

The historic district of St. Charles retains the basic pattern of the town Lewis and Clark saw in 1804. Missouri's first state capitol has been restored, and other buildings from the early nineteenth century serve as shops, homes, and restaurants.

Under the leadership of Glen Bishop, the Discovery Expedition of St. Charles built full-scale replicas of the keelboat and two pirogues used in the 1804 crossing of Missouri. *Darold Jackson.*

The Lewis and Clark Center in St. Charles is devoted entirely to the expedition. Dioramas re-create scenes along the route from Camp Dubois to the Pacific, while other exhibits include authentic artifacts of tribes Lewis and Clark met, a model of the keelboat, and figures representing Sacagawea and other members of the expedition. The center's bookshop has Lewis and Clark titles for every age and interest level.

St. Charles resident Glen Bishop spent a dozen years building by hand a full-scale replica of Lewis and Clark's keelboat that could travel the river. The Discovery Expedition of St. Charles, with Glen Bishop as its guiding spirit, now has full-scale replicas of the two pirogues, as well as a fifty-five-foot keelboat. The boats are not always in St. Charles, however, because their mission is to travel the river and tell the story of Lewis and Clark.

Lewis knew the Femme Osage region was "generally called Boon's settlement" when the expedition moved through in 1804. The Daniel Boone Home, where the elder Boone died in 1820, was built by his son

Holy Family Catholic Church, already standing when Lewis arrived in Cahokia in 1803, was built in the vertical-timber style common in the French Colonial period. The St. Charles church attended by members of the expedition and the St. Louis church where Sacagawea's son was baptized probably looked much like this. *Ann Rogers.*

Nathan in the years Lewis and Clark were making their journey. The attractive, two-story stone house is near Defiance, in St. Charles County, and can be reached by taking Missouri 94 to Route F and following the signs. Boonesfield Village, behind the home, is a collection of buildings, including a steepled church, moved from other sites.

## ILLINOIS

Cahokia, where Lewis first established himself when he arrived in the area, is about three miles from the St. Louis Arch via I-70. Be prepared to exit on Illinois 3 soon after crossing the river into Illinois. Continue south three miles to Holy Family Church, built in 1799, in the French vertical-timber style. The St. Louis church where Sacagawea's son was baptized and the St. Charles church where members of the expedition

Monks Mound, at Cahokia Mounds State Historic Site, was named for the Trappists who established a community at this place for a short period in the nineteenth century. *Cahokia Mounds State Historic Site.*

attended mass would have looked much like this. The Cahokia Court-house Historic Site was built as a residence in 1737, when the region was part of France's colonial empire, and began twenty-one years of service as a courthouse in 1793, with Cahokia under U.S. jurisdiction. The Jarrot Mansion, in the Federal style, was built in 1810 for Nicholas Jarrot, who interpreted for Lewis when he met Spanish officials in St. Louis.

Camp Dubois, where the recruits trained during the winter of 1803–1804, was located on the south bank of Wood River, opposite the Missouri River's entrance into the Mississippi. The Missouri today flows several miles south of its 1804 channel, and the Mississippi has moved east. The wooded site selected for the monument commemorating the winter camp is in the same position relative to the Mississippi and Missouri as the original site, and a nearby visitors center focuses on the preparations for the journey. The Lewis and Clark State Historic Site is two miles south of Hartford, on Illinois 3.

Cahokia Mounds State Historic Site, which is near Collinsville rather than Cahokia, contains within its 2,200 acres sixty-eight man-made mounds used for ceremonies and burials. During his stay at Camp Dubois, Clark saw the northern edges of this group of mounds, which were made by the Mississippian culture, which reached its peak about A.D. 1100. The site includes an excellent interpretive center, level walkways, and a challenge for those who choose to climb the hundred-foot-high Monks Mound, named for the Trappists who lived at this site in the nineteenth century and raised vegetables and fruit trees on the mound's terraces.

Excellent reconstructions of the gatehouse and walls give today's visitors a better perspective on Fort de Chartres than Clark had when he saw it in ruins. About fifty miles southeast of St. Louis, the 1,100-acre Fort de Chartres State Historic Site is four miles west of Prairie du Rocher on Illinois 155. Costumed reenactments throughout the year and an annual rendezvous bring life to the stone fort.

Fort Massac State Park, on the Ohio River at the southern tip of Illinois, features a reconstruction of the 1794 fort Lewis and Clark visited in 1803, a museum, living history weekends, hiking trails, and picnic areas. The Clark statue here is of William's brother George Rogers Clark, who captured the Illinois Territory in the Revolutionary War.


## SOUTHEASTERN MISSOURI

A mural by Jake Wells in the Kent Library of Southeast Missouri State University in Cape Girardeau, about 120 miles south of St. Louis, depicts the early history of the area and includes representations of Tower Rock and Louis Lorimier's house, both visited by Meriwether Lewis. The University Museum houses the Beckwith Collection of pottery and other artifacts of the Mississippian culture. North of town, Cape Rock Park has a circular drive offering panoramic views of the river and a marker telling of Ensign Girardot's arrival.

The legendary Tower Rock, which Lewis scaled and Clark mapped, rises from the Mississippi about one mile south of Wittenberg, in Perry County. Follow Route A east from I-70. The formation is mentioned in the writings of Marquette, Audubon, Mark Twain, and others. The parking and viewing area is unpaved, and there are no walkways or other improvements for today's visitors.

Ste. Genevieve, sixty miles south of St. Louis, is Missouri's oldest permanent settlement. Several buildings that draw visitors today were standing when Clark visited in 1797. The Bolduc House, considered one of the most authentically restored Creole homes in the nation, is typical of the French architecture Lewis and Clark saw in other parts of the Mississippi Valley.

## CENTRAL MISSOURI

Jefferson City, the state capital, is on the Missouri River in a beautiful area of central Missouri. A large relief sculpture depicting the signing of the Louisiana Purchase Treaty stands on the north grounds of the capitol, which overlooks the river. A statue of Thomas Jefferson stands at the building's front entrance. Inside the capitol are statues of Lewis and Clark, a museum of the state's history and natural resources, including a diorama of Fort Osage, and murals showing the explorers' crossing of Missouri and their later visit to President Jefferson.

The Runge Conservation Nature Center, on Missouri 179 at the northwestern edge of town, has a thirty-eight-thousand-gallon aquarium with fish native to Missouri, dioramas of forest, wetland, prairie, and other habitats, a simulated cave that visitors can walk through, a wildlife viewing area equipped with outdoor microphones allowing visitors to hear the sounds of feeding animals, and five hiking trails. As with the Powder Valley Conservation Nature Center in St. Louis, programs and activities are offered for all ages and interests.

Katy Trail State Park, part of a nationwide "rails to trails" movement, offers 225 miles of level, crushed limestone surface for walkers, joggers, and bicyclists. Originally the railroad bed of the Missouri-Kansas-Texas (MKT) Railroad, the trail follows the Missouri River, paralleling Lewis and Clark's route, for much of its distance. From St. Charles westward to Sedalia and, when completed, to Kansas City, there are a number of trailheads and towns offering food and lodging. An especially scenic portion is near Rocheport, where the trail is bordered on one side by the Missouri River and on the other by the bluffs Lewis and Clark described.

Arrow Rock State Historic Site, on Missouri 41, thirteen miles north of I-70 and fourteen miles east of Marshall, takes its name from the "Prairie of Arrows" mentioned in the expedition's journals. Today's visitors are drawn to the shops, museums, craft festivals, and historic buildings. The Old Tavern, built in Clark's lifetime, still serves diners.

Van Meter State Park, off Missouri 41 at Miami, is the ancient home of the Missouri Indians, whose burial mounds remain in the park, along with a six-acre earthwork structure known as the "old fort." Artifacts

The Katy Trail, a hiking and biking trail, parallels much of the expedition's route across Missouri. Near Rocheport the Missouri River is on one side of the trail and bluffs are on the other. *Ann Rogers.*

found on the site and other objects are on display, and a mural in the visitors center provides a view of the people for whom Missouri is named. The beautifully wooded park has facilities for picnics, camping, hiking, and fishing.

## INDEPENDENCE AND KANSAS CITY

The reconstructed Fort Osage, near Sibley, fourteen miles northwest of Independence, holds special events throughout the year, but simply exploring the blockhouses, officers' quarters, barracks, and trading house gives a sense of how the original facility functioned after Clark supervised its construction in 1808. An interpretive center acquaints visitors with the Osage Indians, whom the post served.

In Independence, the National Frontier Trails Center, at 318 West Pacific, has exhibits relating to the Lewis and Clark, Oregon, Santa Fe, and California Trails, with route maps and graphics. Various antiques, such as a spyglass, compass, lap desk, and writing instruments, are similar to those Lewis and Clark used.

A bronze statue of Captains Lewis and Clark, Sacagawea, and York stands at the center of the plaza in Case Park, on Quality Hill, in downtown Kansas City. Eugene Daub's heroic figures overlook the Kansas and Missouri Rivers near the place the captains stood on September 15, 1806, admiring the "Commanding Situation."

The Discovery Center, in Kauffman Legacy Park, near Country Club Plaza, provides city residents and visitors opportunities to increase their outdoor skills and knowledge about nature. A mural by Michael Haynes depicts features of the 1804 landscape, and standing figures of Meriwether Lewis, William Clark, and other expedition members set the stage for the discovery experience. In addition to workshops and exhibits, walkways through various gardens with wildflowers, birds, and butterflies offer a chance to enjoy nature in the heart of Kansas City.

Examples of original and restored prairies can be found in and around Kansas City. The three-hundred-acre Prairie Center, a few miles south of Olathe, Kansas, has big bluestem and a variety of prairie flowers. Jerry Smith Park and the adjacent Saeger Woods Conservation Area have a variety of prairie plants and many birds, including the American woodcock, notable for its courting ritual of spiraling flights and songs. Burr Oak Woods Conservation Area, in Blue Springs, has trails through a variety of natural landscapes, opportunities for viewing wildlife, and an interpretive center similar to the nature centers in St. Louis and Jefferson City.

## KANSAS

Lewis and Clark were army officers leading a military expedition, and in the years that followed the military continued to have a role in America's westward movement. The Frontier Army Museum at Fort Leavenworth has excellent exhibits at the historic post where Custer, Sherman, and Lee were once stationed. York was the first of his race to see the West when he traveled with Lewis and Clark, but he was followed by the black soldiers of the Ninth and Tenth Cavalry, known as the Buffalo Soldiers. A monument to them stands on the grounds.

A shaded overlook with a view of the Missouri River is a feature of Weston Bend State Park. *Ann Rogers.*

Leavenworth has two parks at river level and two esplanades with broad vistas of the river.

U.S. 73 goes north through pleasant farmland to Atchison, which sits on a high bluff overlooking the Missouri. There is a Chamber of Commerce Visitors Center in the Santa Fe Depot, and in front of the depot is a small stream identified as Independence Creek, named by Lewis and Clark on July 4, 1804. At Atchison a bridge crosses the river to Buchanan County, Missouri.

## Northwestern Missouri

Weston Bend State Park, a mile south of Weston on Missouri 45, offers a short biking trail and five miles of hiking trails, but its best-known feature is a picturesque overlook that provides a serene, unspoiled view of river and forests recalling Clark's descriptions of cottonwood and willow along the banks and oak and walnut at higher elevations.

The lake bordering Lewis and Clark State Park may be the one Clark named Gosling Lake when the expedition passed through this area on the Fourth of July in 1804. *Ann Rogers.*

Lewis and Clark State Park, on Missouri 138 in Buchanan County, is opposite Atchison, Kansas, about twenty miles south of St. Joseph, Missouri. The 365-acre lake bordering the park is believed to be the one Clark named Gosling Lake when the expedition passed this site on July 4, 1804. The park has facilities for boating, fishing, picnics, and camping.

The St. Joseph Museum, at 1100 Charles Street, St. Joseph, is housed in a nineteenth-century mansion and features an extensive collection of artifacts illustrating the arts and crafts of the North American Indians, as well as natural history exhibits. A carved image of a Carolina parakeet is part of a diorama relating to extinct birds.

Big Lake State Park, on Missouri 111, about eleven miles south of Mound City, is on a major migratory flyway. Visitors can enjoy birdwatching, fishing, and picnics. Overnight accommodations are available in a motel and cabins in the park, which borders a large oxbow lake formed when the channel of the Missouri River changed.

The expedition's journals contain numerous references to waterfowl and other birds in northwestern Missouri. Squaw Creek National Wildlife Refuge, off U.S. 159, southwest of Mound City, gives photographers and others a chance to see numerous species of migratory birds, as well as beaver, deer, and other wildlife.

Throughout Missouri there are many other reminders of Lewis and Clark and those who joined them on their voyage of discovery. Perhaps their best memorials are the rivers they traveled and the landscapes they described. They saw eight hundred miles of Missouri, traveling its two great rivers and seeing its wetlands and forests and prairies. Again and again they wrote in their journals: "The land is good."

# NOTES

## 1. Preparing for the Journey

1. Donald Jackson, ed., *Letters of the Lewis and Clark Expedition, with Related Documents, 1783–1854,* 1:20.

2. Paul Russell Cutright, "Meriwether Lewis Prepares for a Trip West."

3. Jackson, *Letters,* 1:57–60, 110–11.

4. Ibid., 125.

5. Lewis's account of his Ohio River journey appears in volume 2 of Gary E. Moulton, ed., *The Journals of the Lewis and Clark Expedition.* All quotations from the journals will be from Moulton's edition and will hereafter be cited only if they appear out of sequence and would therefore be difficult to locate.

6. Jackson, *Letters,* 1:58.

7. Moulton, *Journals,* 4:215.

8. Samuel Thomas, "William Clark's 1795 and 1797 Journals and Their Significance."

9. Jackson, *Letters,* 1:7–8.

10. *Missouri: The WPA Guide to the "Show Me" State,* 200.

11. Moulton, *Journals,* 2:113 n. 5.

12. John Shea, *Early Voyages Up and Down the Mississippi,* 68.

13. Moulton, *Journals,* 1:3a, 3b; 2:162.

14. Lewis's descriptions of Missouri in 1804 are found in his "Summary View of the Rivers," in Moulton, *Journals,* 3:336–47. Lewis wrote this document during the winter at Fort Mandan.

15. Thomas, "Clark's 1795 and 1797 Journals," 290.

16. Ibid., 291.

17. Robert Ramsay, *Our Storehouse of Missouri Place Names,* 109.

18. Thomas, "Clark's 1795 and 1797 Journals," 292.

19. Frances Stadler, "St. Louis in 1804," 11–16.

20. Arlen Large, "Expedition Specialists," 5.

21. Moulton, *Journals,* 2:154 n. 3.

22. Pierre Chouteau's father was Pierre Laclède, with whom his mother lived for more than twenty years. Because both French law and the Catholic Church still considered her to be married to the husband who deserted her, her four children by Laclède were given the surname Chouteau (William E. Foley and C. David Rice, *The First Chouteaus: River Barons of Early St. Louis,* 3).

23. Ibid., passim.

24. Jackson, *Letters*, 170–71.

25. Donald Jackson, *Thomas Jefferson and the Stony Mountains: Exploring the West from Monticello*, 161 n. 52.

26. Donald Chaput, "Early Missouri Graduates of West Point."

27. Jackson, *Thomas Jefferson*, 62 n. 7.

28. Robert Hunt, "Matches and Magic."

29. Jackson, *Letters*, 1:192–94.

30. Ibid., 150.

31. James Ronda, *Lewis and Clark among the Indians*, 13, 10.

32. Ibid., 14.

33. Thomas, "Clark's 1795 and 1797 Journals," 292.

34. Moulton, *Journals*, 2:179.

35. Ibid., 159.

36. Jackson, *Letters*, 1:176–77.

37. Ibid., 179, 60, 179.

38. Ibid., 179.

## 2. Westward across Missouri

1. Jo Ann Brown, *History of St. Charles Borromeo*, 44.

2. Moulton, *Journals*, 3:338.

3. The letters *ORD* appear on the cave wall and have been linked by some to John Ordway, although he does not mention visiting the cave.

4. Moulton, *Journals*, 4:262. The term *espontoon*, already becoming obsolete in Lewis's time, referred to a short spear carried by an infantry officer.

5. *Missouri: The WPA Guide*, 364.

6. The known writers were Lewis, Clark, Ordway, Floyd, Gass, and Whitehouse. It is generally assumed that Pryor was writing, because he was a sergeant, and that Frazer was writing, because he later tried to get his journal published.

7. Moulton, *Journals*, 3:338.

8. Ramsay, *Our Storehouse*, 94.

9. Foley and Rice, *First Chouteaus*, 53–55.

10. Jackson, *Letters*, 1:61–62.

11. James Denny, "Lewis and Clark in the Boonslick," 4–7.

12. James Denny, of the Missouri Department of Natural Resources, has identified today's Sugar Loaf Rock in Cole County as the hill Clark described. Gary Moulton, editor of the most recent edition of the Lewis and Clark journals, visited the site and agrees that present-day Sugar Loaf Rock is Clark's Lead Mine Hill of June 4, 1804 (ibid., 8–11).

13. James Wallace, "The Mystery of Clark's Nightingale," 24. Wallace quotes the English visitor but has become convinced the bird was a chuck-will's-widow *(Caprimulgus carolinensis)*.

14. Moulton, *Journals,* 3:452.
15. *Missouri: The WPA Guide,* 359.
16. Ibid., 357.
17. Ibid., 178.
18. Robert Hunt, "The Blood Meal."
19. Moulton, *Journals,* 2:315 n. 11.
20. Joseph Forshaw, *Parrots of the World,* 413.
21. *Missouri: The WPA Guide,* 245–46.
22. Moulton, *Journals,* 2:343 n. 6.
23. Jackson, *Letters,* Biddle Notes, 2:511.
24. Jackson, *Letters,* 1:232.
25. The rodent they saw was likely an eastern wood rat, *Neotoma floridana;* see Charles W. Schwartz and Elizabeth R. Schwartz, *The Wild Mammals of Missouri,* 221–26.
26. Paul Russell Cutright, *Lewis and Clark: Pioneering Naturalists,* 58. Appendix B of the book lists the animals described by Lewis and Clark.
27. Ordway's Missouri creek is possibly Mace Creek, north of the Andrew-Buchanan county line (Moulton, *Journals,* 2:360 n. 2, 9:22 n. 1).
28. Ibid., 2:363 nn. 3, 9.

### 3. The Months Between

1. Lewis and Clark's Council Bluff was on the west side of the Missouri River, in present-day Nebraska, and north of today's city of Council Bluffs, Iowa.
2. For a description of the military and religious ceremonies that would probably have taken place, see Robert Hunt, "For Whom the Guns Sounded."
3. Jackson, *Letters,* 1:370 n. 3. George Yater identifies the letter's writer, Nathaniel Floyd, and its recipient, Nancy, as cousins of Sergeant Floyd ("Nine Young Men from Kentucky," 5–6). Yater blames the confusion as to whether they were his cousins or his siblings on the fact that their father was named Charles Floyd. The elder Floyd, Yater says, was the sergeant's uncle rather than his father.
4. Eldon G. Chuinard, *Only One Man Died: The Medical Aspects of the Lewis and Clark Expedition,* 238–39.
5. Jackson, *Letters,* 1:254–56.
6. Sacagawea's year of birth is not known; she was probably about sixteen or seventeen in 1805.
7. Moulton, *Journals,* 4:175 n. 5.
8. They named the others the Madison, for Secretary of State James Madison, and the Gallatin, for Secretary of the Treasury Albert Gallatin.
9. Moulton, *Journals,* 5:76 n. 8.
10. Jackson, *Letters,* 2:519.

11. Roy Appleman, *Lewis and Clark: Historic Places Associated with Their Transcontinental Exploration,* 373 n. 120.

12. York was the first of his race to achieve these goals.

13. Wapatoo or wappato, an arrowhead plant (in the genus *Sagittaria*) with edible tubers, was not found near the expedition's fort; it grew in swampy places and was eaten by the Indians of the Northwest. "The bulb of the plant . . . when roasted, tasted much like a potato" (Cutright, *Lewis and Clark: Pioneering Naturalists,* 265).

## 4. The Return through Missouri

1. The Spanish had made at least three unsuccessful attempts to intercept the expedition (Jackson, *Thomas Jefferson,* 153–54).

2. Ken Walcheck, "Wapiti," 31.

3. Moulton, *Journals,* 2:306.

4. Chuinard, *Only One Man Died,* 395.

5. Jan Phillips, *Wild Edibles of Missouri,* 217; Walter Muenscher, *Poisonous Plants of the United States,* 7, 98; see also John Tampion, *Dangerous Plants,* 164.

6. Moulton, *Journals,* 3:453.

7. Volume 12 of Moulton's edition of the Lewis and Clark journals is devoted to the plant specimens Lewis collected.

## 5. Missouri Sequels

1. Jackson, *Letters,* 1:319–23.

2. Ibid., 323–24.

3. Foley and Rice, *First Chouteaus,* 114.

4. Jackson, *Letters,* 2:441.

5. Richard H. Dillon, *Meriwether Lewis: A Biography,* 290.

6. Ruth Colter Frick, "Conflict: Frederick Bates and Meriwether Lewis," 20.

7. William E. Foley and C. David Rice, "The Return of the Mandan Chief."

8. Ibid., 10.

9. James Holmberg, "I Wish You to See & Know All," 10.

10. Ibid., 11. Lewis's servant, traveling with him, was the source of this information.

11. In the period leading up to his death, Lewis was apparently suffering from depression, abusing alcohol, and experiencing recurrences of malaria he had contracted on the Ohio River in 1803. He was unsuited by temperament to govern the Louisiana Territory and was agitated by the charges brought against him. The two persons who knew him best in his adult years, William Clark and Thomas Jefferson, believed he took his own life.

12. James Holmberg, "Seaman's Fate," 8.

13. Ibid., 8–9.

14. Clark kept a daily record except during a brief hunting trip at the Mandans, and he summarized those days' events on his return (Paul Russell Cutright, *A History of the Lewis and Clark Journals*, 9).

15. James Lal Penick Jr., *The New Madrid Earthquakes, Revised Edition.*

16. Timothy Flint, quoted in ibid., 50.

17. Larry Morris, "Dependable John Ordway," 32.

18. Charles G. Clarke, *The Men of the Lewis and Clark Expedition: A Biographical Roster of the Fifty-one Members and a Composite Diary of Their Activities from All the Known Sources*, 46.

19. Moulton, *Journals*, 8:302.

20. "Colter's Hell" originally applied to a region he saw east of today's Yellowstone National Park, but the name has sometimes been applied to the Yellowstone area, with its mudpots, geysers, and boiling springs.

21. Burton Harris, *John Colter: His Years in the Rockies*, 158, 128–31. Colter's escape is an often told story, and details are different in the various accounts.

22. Ibid., 162–63. Ruth Frick, a Colter descendant, believes he was buried next to his son, on a farm near New Haven, Missouri.

23. M. O. Skarsten, *George Drouillard: Hunter and Interpreter for Lewis and Clark and Fur Trader, 1807–1810*, 20.

24. Richard Oglesby, *Manuel Lisa and the Opening of the Missouri Fur Trade*, 41.

25. Jackson, *Letters*, 1:368; Skarsten, *George Drouillard*, 271–79.

26. Robert Lange, "Private George Shannon," 13.

27. Jackson, *Letters*, 2:569.

28. Ibid., 2:620–21 n. 2.

29. John McBride, "Pioneer Days in the Mountains," 316.

30. Jackson, *Letters*, 1:345–46.

31. Moulton, *Journals*, vol. 1, map 124.

32. Cutright, *History.*

33. Clarke, *Men of the Expedition*, 53.

34. Jackson, *Letters*, 1:367.

35. Clarke, *Men of the Expedition*, 54.

36. Ibid., 43–45.

37. Jackson, *Letters*, 1:347.

38. Clarke, *Men of the Expedition*, 49.

39. Jackson, *Letters*, 1:367.

40. Clarke, *Men of the Expedition*, 64.

41. Ibid., 56.

42. Clarke's *Men of the Expedition* has a brief biographical sketch of each man.

43. Moulton, *Journals*, 3:157, 9:85, 3:157.

44. Robert B. Betts, *In Search of York: The Slave Who Went to the Pacific*

*with Lewis and Clark,* 106, notes that "nowhere in any of the journals . . . is there a derogatory remark made about him [York] or is an act suggesting racial prejudice reported."

45. Holmberg, "I Wish You to See & Know All," 7.

46. Betts, *In Search of York,* 112–13.

47. Ibid., 108.

48. Ibid., 119.

49. Clarke, *Men of the Expedition,* 60.

50. Ibid.

51. Ibid., 65–70; Moulton, *Journals,* 2:525–29; Jo Ann Brown, "New Light on Some of the Expedition Engages."

52. Jackson, *Letters,* 1:315–16.

53. Bob Moore, "Pompey's Baptism." The first three children born to William and Julia Clark were baptized in this same church on August 8, 1814, by Bishop Benedict Flaget of Kentucky during his visit to St. Louis that year. I thank Monsignor Bernard Sandheimrich for allowing me to examine the Old Cathedral's Register of Baptisms for 1814.

54. Ibid., 11.

55. Ibid., 11–14.

56. H. M. Brackenridge, *Journal of a Voyage Up the River Missouri, Performed in Eighteen Hundred and Eleven,* 32–33. Brackenridge added that Sacagawea's husband "had become weary of a civilized life."

57. John C. Luttig, *Journal of a Fur-Trading Expedition on the Upper Missouri, 1812–1813,* 106. Another account, unaccepted by most historians, has Sacagawea leaving Charbonneau, eventually finding her way to Shoshones in Wyoming, and dying there in 1884. This version first appeared before Luttig's journal was published and before the discovery of Clark's notations from the 1820s in which he listed Sacagawea as dead.

58. Ibid., 15.

59. Harold P. Howard, *Sacajawea,* 161.

60. Luttig, *Journal,* 140–41.

61. Ann Hafen, "Jean Baptiste Charbonneau."

62. Kate Gregg, ed., *Westward with Dragoons: The Journal of William Clark on His Expedition to Establish Fort Osage, August 25 to September 22, 1808.*

63. When Meriwether Lewis returned to St. Louis to assume his duties as governor, his brother accompanied him. After serving at Fort Osage and as a partner in the Missouri Fur Company, he returned to Virginia, where he died in 1844.

64. Company Book of John Symmes and Eli Clemson, National Archives RG 98. The date of Whitehouse's birth as ascertained from this record does not correspond to that in other sources, but it seems likely he was the former expedition member. I am indebted to Dave Bennett, historian at Fort Osage, for the information and regret I have not yet reconciled the differences between this record and others.

65. Clark to Dearborn, June 25, 1808, and Lewis to Dearborn, July 1,

1808, *Territorial Papers of the United States*, 14:194–203, Clarence Carter, ed.

66. Elliott Coues, ed., *History of the Expedition under the Command of Lewis and Clark*, 1:30.

67. Jackson, *Letters*, Biddle Notes, 2:509.

68. The map is reproduced in volume 1 of Moulton's edition of the *Journals*.

69. Holmberg, "I Wish You to See & Know All," 11.

70. Jackson, *Letters*, Clark Memorandum, 2:486.

71. Cutright, *History*, 62–66.

72. Robert C. Carriker, *Father Peter John De Smet: Jesuit in the West*, 9, 39.

73. Louise Callan, *Philippine Duchesne*, 358. By 1822 Clark would have been the former governor of the Missouri Territory.

74. Ibid., 271.

75. John Francis McDermott, "Museums in Early St. Louis," and John Ewers, "William Clark's Indian Museum in St. Louis, 1816–1838." Clark's home was torn down in 1851 and replaced with warehouses and eventually by the grounds of the Jefferson National Expansion Memorial. No homes from Clark's period survive in downtown St. Louis.

76. Ewers, "William Clark's Indian Museum," 66–70; John Francis McDermott, "William Clark's Museum Once More."

77. Jackson, *Letters*, 638–39. Clark's listing of Lewis as "dead" rather than "killed" is evidence that Clark believed Lewis's wounds were self-inflicted.

78. Donald Jackson, *Voyages of the Steamboat* Yellow Stone, 31–35.

79. Bernard DeVoto, ed., *The Journals of Lewis and Clark*, xlviii.

80. Ibid., xlix.

## 6. Lewis and Clark and the 1904 Louisiana Purchase Exposition

1. *The Forest City: Official Photographic Views of the Universal Exposition*, 116.

2. Mark Bennitt, ed., *History of the Louisiana Purchase Exposition*, 73.

3. Robert Lange, "1905 Statues of the Captains Have Disappeared," 20.

4. Karl Bitter, "Sculpture of the Louisiana Purchase Exposition," in David R. Francis, *The Universal Exposition of 1904*, 44.

5. "Louisiana Purchase Exposition Sculpture and Sculptors" (Missouri Historical Society vertical file).

6. *World's Fair Bulletin* (September 1903), 31.

7. The University of Virginia's *College Topics*, October 12, 1904, reported the shipment of the Galt statue to St. Louis.

8. William Peden, "The Jefferson Monument at the University of Missouri."

9. Cutright, *History*, 118–21.

10. Howard Obear, *The Louisiana Purchase Exposition*, 29, 26.

11. June Dahl, *A History of Kirkwood, Missouri, 1881–1965*, 328.

12. Obear, *Louisiana Purchase Exposition*, 25.

13. *St. Louis Republic*, October 3, 1904, p. 11.

14. Ibid.

15. Cutright, *History*, 124–25.

16. Francis, *Universal Exposition*, 679.

17. A list of the society's major Lewis and Clark holdings appears in Cutright's *Lewis and Clark: Pioneering Naturalists*, 452–53. Jan Snow, "Lewis and Clark in the Museum's Collections," has color photographs of the telescope, compass, and related items.

# BIBLIOGRAPHY

Appleman, Roy. *Lewis and Clark: Historic Places Associated with Their Transcontinental Exploration.* Washington, D.C.: National Park Service, 1975.

Bennitt, Mark, ed. *History of the Louisiana Purchase Exposition.* St. Louis: Universal Exposition Publishing, 1905.

Betts, Robert B. *In Search of York: The Slave Who Went to the Pacific with Lewis and Clark.* Boulder: Colorado Associated University Press, 1985.

Brackenridge, H. M. *Journal of a Voyage Up the River Missouri, Performed in Eighteen Hundred and Eleven.* In *Early Western Travels, 1748–1846,* ed. Reuben Gold Thwaites. Vol. 6. Cleveland: Arthur H. Clark, 1904.

Bradbury, John. "Travels in the Interior of America, in the Years 1809, 1810, and 1811." In *Early Western Travels, 1748–1846,* ed. Reuben Gold Thwaites. Vol. 5. Cleveland: Arthur H. Clark, 1904.

Brown, Jo Ann. *A History of St. Charles Borromeo.* St. Louis: Patrice Press, 1991.

———. "New Light on Some of the Expedition Engages." *We Proceeded On* 22, no. 3 (August 1996): 14–19.

Callan, Louise. *Philippine Duchesne: Frontier Missionary of the Sacred Heart, 1769–1852.* Westminster, Md.: Newman Press, 1957.

Carriker, Robert C. *Father Peter John De Smet: Jesuit in the West.* Norman: University of Oklahoma Press, 1995.

Carter, Clarence, ed. *Territorial Papers of the United States.* Vol. 14. Washington, D.C.: U.S. Government Printing Office, 1934.

Chaput, Donald. "Early Missouri Graduates of West Point." *Missouri Historical Review* 72, no. 3 (April 1978): 262–70.

Chuinard, Eldon G. *Only One Man Died: The Medical Aspects of the Lewis and Clark Expedition.* Glendale, Calif.: Arthur H. Clark, 1979.

Clarke, Charles G. *The Men of the Lewis and Clark Expedition: A Biographical Roster of the Fifty-one Members and a Composite Diary of Their Activities from All the Known Sources.* Glendale, Calif.: Arthur H. Clark, 1970.

Coues, Elliott, ed. *History of the Expedition under the Command of Lewis and Clark.* 1893. Reprint, New York: Dover Publications, 1965.

Cutright, Paul Russell. *A History of the Lewis and Clark Journals.* Norman: University of Oklahoma Press, 1976.

———. *Lewis and Clark: Pioneering Naturalists.* Lincoln: University of Nebraska Press, 1969.

———. "Meriwether Lewis Prepares for a Trip West." *Missouri Historical Society Bulletin* 23, no. 1 (October 1966): 3–19.

Dahl, June. *A History of Kirkwood, 1881–1965.* Kirkwood, Mo.: Kirkwood Historical Society, 1965.

Denny, James. "Lewis and Clark in the Boonslick." *Boonslick Heritage* 8, nos. 2–3 (June–September 2000): 3–27.

DeVoto, Bernard. *Across the Wide Missouri.* Boston: Houghton Mifflin, 1975.

DeVoto, Bernard, ed. *The Journals of Lewis and Clark.* Boston: Houghton Mifflin, 1953.

Dillon, Richard H. *Meriwether Lewis: A Biography.* New York: Coward-McCann, 1965.

Ewers, John. "William Clark's Indian Museum in St. Louis, 1816–1838." In *A Cabinet of Curiosities: Five Episodes in the Evolution of American Museums,* by Whitfield J. Bell et al., 49–72. Charlottesville: University Press of Virginia, 1967.

Foley, William E., and C. David Rice. *The First Chouteaus: River Barons of Early St. Louis.* Urbana: University of Illinois Press, 1983.

———. "The Return of the Mandan Chief." *Montana: The Magazine of Western History* 29, no. 3 (summer 1979): 2–14.

*The Forest City: Official Photographic Views of the Universal Exposition.* St. Louis: N. D. Thompson Publishing, 1904.

Forshaw, Joseph. *Parrots of the World.* Illustrated by William Cooper. Melbourne, Australia: Lansdowne Press, 1973; Neptune, N.J.: T.F.H. Publications, 1977.

Francis, David R. *The Universal Exposition of 1904.* St. Louis: Louisiana Purchase Exposition Co., 1913.

Frick, Ruth Colter. "Conflict: Frederick Bates and Meriwether Lewis." *We Proceeded On* 19, no. 3 (August 1993): 20–24.

Gregg, Kate, ed. *Westward with Dragoons: The Journal of William Clark on His Expedition to Establish Fort Osage, August 25 to September 22, 1808.* Fulton, Mo.: Ovid Bell Press, 1937.

Hafen, Ann. "Jean Baptiste Charbonneau." In *The Mountain Men and the Fur Trade of the Far West,* vol. 1, ed. LeRoy R. Hafen, 205–24.

Glendale, Calif.: Arthur H. Clark, 1965.

Harris, Burton. *John Colter: His Years in the Rockies.* 1952. Reprint, Lincoln: University of Nebraska Press, 1993.

Holmberg, James. "I Wish You to See & Know All." *We Proceeded On* 18, no. 4 (November 1992): 4–11.

———. "Seaman's Fate." *We Proceeded On* 26, no. 1 (February 2000): 7–9.

Howard, Harold P. *Sacajawea.* Norman: University of Oklahoma Press, 1971.

Hunt, Robert. "The Blood Meal." *We Proceeded On* 18, nos. 2–3 (May–August 1992): 4–10.

———. "For Whom the Guns Sounded." *We Proceeded On* 27, no. 1 (February 2001): 10–15.

———. "Matches and Magic." *We Proceeded On* 26, no. 3 (August 2000): 13–17.

Jackson, Donald. *Thomas Jefferson and the Stony Mountains: Exploring the West from Monticello.* Norman: University of Oklahoma Press, 1993.

———. *Voyages of the Steamboat* Yellow Stone. Norman: University of Oklahoma Press, 1985.

Jackson, Donald, ed. *Letters of the Lewis and Clark Expedition, with Related Documents, 1783–1854.* 2d ed., 2 vols. Urbana: University of Illinois Press, 1978.

Lange, Robert. "1905 Statues of Captains Have Disappeared." *We Proceeded On* 6, no. 4 (November 1980): 20.

———. "Private George Shannon." *We Proceeded On* 8, no. 3 (July 1982): 10–15.

———. "William Bratton." *We Proceeded On* 7, no. 1 (February 1981): 8–11.

Large, Arlen. "Expedition Specialists." *We Proceeded On* 20, no. 1 (February 1994): 4–10.

Luttig, John C. *Journal of a Fur-Trading Expedition on the Upper Missouri, 1812–1813.* Ed. Stella M. Drumm. New York: Argosy-Antiquarian, 1964.

McBride, John. "Pioneer Days in the Mountains." *Tullidge's Quarterly Magazine of Utah* 3 (July 1884): 311–20.

McDermott, John Francis. "Museums in Early St. Louis." *Missouri Historical Society Bulletin* 4, no. 3 (April 1948): 129–38.

———. "William Clark's Museum Once More." *Missouri Historical Society Bulletin* 16, no. 2 (January 1960): 130–33.

*Missouri: The WPA Guide to the "Show-Me" State.* 1941. Reprint, St. Louis: Missouri Historical Society Press, 1998.

Moore, Bob. "Pompey's Baptism." *We Proceeded On* 26, no. 1 (February 2000): 10–17.

Morris, Larry. "Dependable John Ordway." *We Proceeded On* 27, no. 2 (May 2001): 28–33.

Moulton, Gary E., ed. *The Journals of the Lewis and Clark Expedition.* 13 vols. Lincoln: University of Nebraska Press, 1986–2001.

Muenscher, Walter. *Poisonous Plants of the United States.* New York: Macmillan, 1961.

Obear, Howard. *The Louisiana Purchase Exposition.* Chicago: Cable Company, 1904.

Oglesby, Richard. *Manuel Lisa and the Opening of the Missouri Fur Trade.* Norman: University of Oklahoma Press, 1963.

Peden, William. "The Jefferson Monument at the University of Missouri." *Missouri Historical Review* 72, no. 1 (October 1977): 67–77.

Penick, James Lal, Jr. *The New Madrid Earthquakes, Revised Edition.* Columbia: University of Missouri Press, 1981.

Phillips, Jan. *Wild Edibles of Missouri.* Jefferson City: Missouri Department of Conservation, 1979.

Ramsay, Robert. *Our Storehouse of Missouri Place Names.* Columbia: University of Missouri Press, 1973.

Ronda, James. *Lewis and Clark among the Indians.* Lincoln: University of Nebraska Press, 1984.

*St. Louis Republic,* October 3, 1904.

Schwartz, Charles W., and Elizabeth R. Schwartz. *The Wild Mammals of Missouri, Second Revised Edition.* Columbia: University of Missouri Press; Jefferson City: Missouri Department of Conservation, 2001.

Shea, John. *Early Voyages Up and Down the Mississippi.* Albany: J. Munsell, 1861.

Skarsten, M. O. *George Drouillard: Hunter and Interpreter for Lewis and Clark and Fur Trader, 1807–1810.* Glendale, Calif.: Arthur H. Clark, 1964.

Snow, Jan. "Lewis and Clark in the Museum's Collections." *Gateway Heritage* 2, no. 2 (fall 1981): 36–41.

Stadler, Frances. "St. Louis in 1804." *We Proceeded On* 20, no. 1 (February 1994): 11–16.

Tampion, John. *Dangerous Plants.* New York: Universe Books, 1977.

Thomas, Samuel. "William Clark's 1795 and 1797 Journals and Their Significance." *Missouri Historical Society Bulletin* 25, no. 4 (July 1969): 277–95.

Walcheck, Ken. "Wapati." *We Proceeded On* 26, no. 3 (August 2000): 26–32.

Wallace, James. "The Mystery of Clark's Nightingale." *We Proceeded On* 26, no. 2 (May 2000): 18–25.

Yater, George. "Nine Young Men from Kentucky." *We Proceeded On* supplementary publication, no. 11 (May 1992): 1–14.

# INDEX

# About the Author

Ann Rogers is a member of the Missouri Lewis and Clark Bicentennial Commission. She resides in St. Louis, Missouri.